SOLID MECHANICS
ADVANCED STUDY

SOLID MECHANICS
ADVANCED STUDY

Dr. Tawfiq Abdul-Aziz Al-Mughanam
Dr. Vineet Tirth
Dr. Ali Algahtani
Dr. Abdulaziz Hossain Mohammed Alghtani

NAMYA PRESS.com

Published by

Namya Press

India: 213, Vardan House, 7/28 Ansari Road Darya Ganj - Delhi India 110002
USA: 10685-B Hazelhurst Dr. # 31013 Houston, Texas USA - 77043
Email: namyapress@gmail.com
Website: https://namyapress.com

Edition	:	First Published in 2023 (Namya Press)
Title	:	Solid Mechanics: Advanced Study
Authors	:	Dr. Tawfiq Abdul-Aziz Al-Mughanam, Dr. Vineet Tirth, Dr. Ali Algahtani & Dr. Abdulaziz Hossain Mohammed Alghtani
ISBN	:	978-1-956861-31-0
Copyright	:	© Dr. Tawfiq Abdul-Aziz Al-Mughanam, Dr. Vineet Tirth, Dr. Ali Algahtani & Dr. Abdulaziz Hossain Mohammed Alghtani 2023 All Rights Reserved

Preface/Acknowledgement

The authors acknowledge the Deanship of Scientific Research, Vice Presidency for Graduate Studies and Scientific Research at King Faisal University, Saudi Arabia for financial support under the annual funding track [GRANT3439].

The authors extend their appreciation to the College of Engineering at King Khalid University Abha 61421, Asir, Kingdom of Saudi Arabia for providing support and facilities to complete this book.

Author Contribution

All authors contributed equally in the book in conceptualization, writing, review and editing.

Contents

1

Introduction: Stress and Strain of Solid Mechanics

ABSTRACT

The investigation of anxiety is central to the field of strong mechanics. When designing and analyzing various structures and mechanical components, it is essential to comprehend how materials respond to external forces and deform under stress. The definitions of stress and strain, as well as their connection to solid mechanics, are discussed in this introduction. Plastic deformation, elastic deformation, and the mechanical properties of materials are all discussed. It also emphasizes the significance of accurate modeling and prediction of material behavior under various loading conditions and the significance of stress and strain analysis in engineering applications.

Keywords: *deformation, mechanical properties, elastic deformation, plastic deformation, engineering applications, material behavior, and loading conditions are all examples of stress and strain.*

INTRODUCTION

In the field of strong mechanics, anxiety is a major idea that plays a significant role in figuring out the way materials behave under outer burdens. Engineers and scientists can design safe and effective systems by analyzing and anticipating the response of structures and components to various forces through the study of stress and strain.

A material's internal resistance or force when subjected to external loads is referred to as stress. It is a measure of the strength of the forces that are present within the material, and it has the potential to alter the shape and size of an object as well as cause deformation. Strain, then again, is the proportion of disfigurement or lengthening experienced by a material under pressure. It gives a number to the relative change in the material's size or shape caused by forces.

In numerous engineering fields, such as civil, mechanical, aerospace, and materials engineering, understanding stress and strain is essential. To guarantee the durability and stability of structures, engineers must examine how stress and strain are distributed within them. Engineers can determine the design requirements and select suitable materials to withstand the anticipated loads by assessing the levels of stress and strain.

Particles within a material undergo deformation when subjected to external forces. The manner in which a material answers these powers depends upon its mechanical properties, like flexibility, pliancy, and strength. These properties determine the material's behavior under various stress levels and its ultimate resistance to deformation or failure.

The capacity of a material to deform under stress and return to its original shape when the stress is removed is referred to as its elasticity. Hooke's law states that a material will have a linear relationship between stress and strain if it behaves elastically. Engineers can use this relationship to predict the behavior of a structure and calculate its stress and strain.

In any case, when the weight of a material surpasses its versatile breaking point, it experiences plastic deformity. Plastic distortion is non-reversible, implying that the material doesn't get back to its unique shape after the pressure is removed. Pliancy is a significant consideration in the planning of designs, as over-the-top plastic distortion can prompt disappointment and compromise the wellbeing of the framework.

Anxiety investigation is performed through exploratory testing, logical computations, and mathematical reproductions. Material samples are frequently subjected to mechanical tests by engineers to ascertain their stress-strain behavior. These tests include applying controlled powers or loads to the example and estimating the subsequent distortion. By plotting the pressure strain bend, specialists can portray the material's way of behaving and remove significant mechanical properties like Young's modulus, yield strength, and extreme rigidity.

Stress and strain distributions in complex structures can also be predicted using analytical calculations and mathematical models. Equations derived from the fundamental laws of physics and solid mechanics are used in these calculations. Specialists can apply these conditions to survey the anxiety in various parts of a framework, empowering them to enhance plans and guarantee underlying respectability.

For stress and strain analysis, numerical simulations and computer-aided engineering (CAE) tools have become essential in recent years. Finite Element Analysis (FEA) is a generally utilized mathematical strategy that discretizes complex designs into more modest components to re-enact their conduct under different burdens. Before physical prototyping, engineers can use FEA to visualize stress and strain distributions, identify potential failure points, and optimize designs, reducing development costs and time.

The investigation of the anxiety of strong mechanics isn't simply restricted to designing applications; it also tracks down significance in other logical fields. Materials researchers explore the ways in which various materials behave under different stacking conditions to work on their presentation and foster new materials with improved properties. Additionally, biomechanics researchers aid in the development of prosthetics, implants, and rehabilitation strategies by examining the stress and strain in biological tissues to comprehend the human body's mechanics.

Solid mechanics' stress and strain analysis is an essential component of engineering and scientific investigation. By understanding the way materials behave under outer burdens, architects can plan protected and effective designs, while researchers can foster new materials with further developed properties. To guarantee the performance and dependability of various systems, it is essential to be able to accurately predict stress and strain distributions using experimental, analytical, and numerical techniques. With continuous progressions in innovation and computational apparatuses, the investigation of anxiety keeps on advancing, empowering more precise re-enactments and working with creative design arrangements.

Background

It is essential to investigate the fundamental concepts and theories that govern solid mechanics to delve deeper into stress and strain. The behavior of materials at both the macro and microscales is examined in the study of stress and strain. Mathematical equations and principles are used to describe the material's behavior at the macroscale, where it is viewed as a continuum. To comprehend where stress and strain originate, the material's atomic and molecular interactions at the microscopic level are taken into consideration.

The way a material behaves under pressure can be sorted into two fundamental sorts: deformation in both elastic and plastic Flexible disfigurement is reversible, implying that the material re-visits its unique shape once the applied powers are removed. Plastic deformation, on the other hand, is irreversible and causes the shape of the material to change forever. When the material's stress exceeds a certain threshold, known as the yield strength, it moves from elastic to plastic deformation.

To evaluate anxiety, a few boundaries are ordinarily utilized. Stress is commonly addressed by the image (sigma) and is communicated in units of power per unit region, like Pascal or pounds per square inch (psi). Dimensionless strain is denoted by the symbol "epsilon. It can be expressed as the ratio of the material's original length to the length that has changed or deformed.

One of the central connections in strong mechanics is Hooke's regulation, which portrays the straight connection between anxiety and flexible deformity. Within the material's elastic limit, stress is directly proportional to strain, according to Hooke's law. The mathematical expression for this relationship is "= E," where E is the material's elastic modulus or Young's modulus. The material's stiffness and resistance to deformation are measured by its elastic modulus.

The mechanical properties and responses to stress and strain of various materials can vary significantly. Ceramics, for instance, are brittle materials with a tendency to have a low elastic modulus and easily fracture under stress. Pliable materials, like metals, have a higher flexible modulus and can go through significant plastic twisting before disappointing. When choosing materials for specific applications and designing structures that can withstand anticipated loads, having an understanding of these material properties is essential.

Materials may exhibit nonlinear behavior under large deformations or high stresses in addition to the linear relationship that is described by Hooke's law. More intricate mathematical models, such as the nonlinear stress-strain curves for plastic deformation, can be used to explain this behavior. The material's strain hardening, or increase in strength and resistance to deformation as plastic deformation progresses, is captured in these curves.

Additionally, depending on the direction of the forces applied and the material's response, stress and strain can have distinct components. Compressive stress and strain, on the other hand, occur when forces act to compress or shorten a material, whereas tensile stress and strain occur when forces act to elongate a material. When forces act in opposite directions but parallel to one another, sliding or deformation of the material's layers occurs. This results in shear stress and strain.

Objectives

The behavior of solid materials under a variety of external forces and deformations is the subject of solid mechanics, a fundamental engineering and physics field. The study of stress and strain, which provides insights into the response of materials to applied loads, is one of the most important aspects of solid mechanics. The design and analysis of structures, as well as the prediction of materials' failure, all depend on having a clear understanding of the relationship between stress and strain. The purpose of this study is to investigate the goals of studying stress and strain in solid mechanics, highlighting their significance and potential uses.

1. How Material Behaviour Is Defined: The essential target of this exploration is to describe the way materials behave under various stacking conditions. The response of materials to external forces and deformations can be better understood through stress and strain analysis. Researchers can ascertain a material's mechanical properties, such as its stiffness, strength, and ductility, by quantifying the distribution of stress and strain within the material. Choosing the right materials for a given application, optimizing designs, and ensuring structural integrity all depend on this knowledge.

2. Predicting Failure in the Structure: The prediction and prevention of structural failures is an additional important goal of studying stress and strain in solid mechanics. Researchers can determine the design's safety margin and potential failure points by examining the stress distribution within a structure. When determining whether a material will fracture or undergo plastic deformation under specific loading conditions, it is helpful to comprehend the strain levels and their distribution. Engineers can make well-informed choices regarding the design of structural elements, the selection of materials, and the implementation of safety measures thanks to this knowledge.

3. Optimization of Design: Proficient plan improvement is a key objective in strong mechanics examination, and anxiety investigation plays a critical role in accomplishing this objective. Researchers can identify regions of high-stress concentration or excessive deformation by studying the patterns of stress and strain within a structure. By redistributing the stresses, altering material selections, or altering the structural geometry, this information aids in design enhancement. Using stress and strain analysis to improve designs can result in lighter, more energy-efficient structures with lower material consumption and associated costs while maintaining or increasing performance.

4. Material Choice and Advancement: In solid mechanics, research on stress and strain aids in the selection and creation of new materials with improved mechanical properties. Researchers can identify materials with superior mechanical properties for specific applications by comparing their performance characteristics and studying the stress-strain behavior of various materials. Advanced materials with enhanced strength, ductility, fatigue resistance, and other desirable properties can be developed more easily with this knowledge. Additionally, it assists in the selection of materials for particular engineering applications based on how they respond to various loading conditions.

5. Theoretical Model and Numerical Simulation Validation: Approving hypothetical models and mathematical re-enactments is a fundamental goal of anxiety research. By contrasting trial results with hypothetical expectations or recreation yields, scientists can assess the precision and unwavering quality of the models utilized. This cycle helps in refining and working on existing hypotheses and computational calculations, prompting more precise forecasts of anxiety behavior. Precise displaying and recreation procedures are pivotal for designing plans, permitting specialists to anticipate the exhibition of designs and materials without costly and tedious exploratory testing.

Scope

1. **Central Ideas of Anxiety:** The fundamental concepts of stress and strain in solid mechanics will be the focus of the research's initial investigation. This will include a thorough investigation of the definitions, units, and numerical portrayals of anxiety. Various sorts of pressure, for example, typical pressure, shear pressure, and chief pressure, will be investigated, along with relating strain measures.

2. **Flexibility and Hooke's Regulation:** The examination will encompass a top-to-bottom investigation of the versatile way materials behaves and their relationship with anxiety. Hooke's Regulation, which relates anxiety through the modulus of versatility, will be analyzed. Stress-strain curves, linear and nonlinear elastic materials, and experimentally determining material properties are all parts of the investigation.

3. **Fracture mechanics and concentration of stress:** The review will zero in on pressure-focus peculiarities and their effect on the strength and unwavering quality of designs. It will encompass an investigation of stress focus factors, the job of mathematical highlights in pressure fixation, and the related disappointment systems. To comprehend how materials behave under extreme loading conditions, we will investigate fracture mechanics concepts like crack propagation and fracture toughness.

4. **Criteria for Plasticity and Yield:** The exploration will dig into the plastic way materials behave, which happens when feelings of anxiety surpass as far as possible. The review will incorporate plastic disfigurement, strain solidifying, and the idea of yield rules. The applicability of various yield criteria, such as Von Mises, Tresca, and Mohr-Coulomb, to various materials and loading scenarios, will be investigated.

5. **Analyses of fatigue and creep:** The phenomena of creep and fatigue, which are crucial in assessing a material's durability and long-term performance, will also be covered in this study. Creep alludes to the time-subordinate twisting of materials under consistent pressure, while weakness includes the disappointment of materials exposed to cyclic stacking. The review will investigate the systems, life expectation models, and moderation methodologies for both jerks and weaknesses.

6. Finite Element Analysis (FEA) in Anxiety Examination: The application of numerical techniques to stress and strain analysis, particularly finite element analysis (FEA), will also be part of the scope of the research. The study will investigate the principles and ways in which FEA can be used to solve difficult problems in solid mechanics. It will discuss the generation of meshes, boundary conditions, material modeling, and the interpretation of results.

Methods for Experimentation and Validation

The research will incorporate experimental methods for measuring stress and strain to verify the theoretical analysis. Different trial strategies, for example, strain checks, extensometers, and optical methods will be investigated. Understanding the limitations and accuracy of experimental measurements, as well as the relationship between numerical and experimental results, will be the primary focus.

Limitations or Assumptions

This exposition investigates the different constraints and presumptions related to the prologue to anxiety in strong mechanics, revealing insight into their suggestions and expected influence on designing applications.

1. Isotropic and homogeneous materials: The assumption of materials' homogeneity and isotropy is one of the primary assumptions used in stress and strain analysis. It assumes that materials' mechanical properties are consistent across all directions and have the same responses. Notwithstanding, actually, most materials show a few levels ofheterogeneity and anisotropy, which can fundamentally influence their mechanical way of behaving. Inadequate predictions and compromised structural designs may result if these deviations are not taken into account.

2. Elasticity Linear: One more generally embraced supposition in strong mechanics is the presumption of straight versatility, which suggests that the connection among anxiety is direct inside the flexible reach. However, the linear elastic assumption is insufficient because many real-world materials behave non-linearly when subjected to high loads or prolonged loading periods. Predicting material failure and deformation may be inaccurate if non-linear effects are not taken into account.

3. Minor Deviations: Small deformations, in which a solid body's shape or size changes relatively little in comparison to its initial dimensions, are commonly used as the starting point for the introduction of stress and strain. The application of linear elastic models is made easier by this assumption, which makes mathematical analysis simpler. However, it restricts the theory's applicability to situations involving significant deformations, such as nonlinear material behavior, large-scale structural deformations, or plasticity.

4. Stress and strain on the plane: The introduction to stress and strain frequently assumes either plane stress or plane strain conditions to simplify the analysis. Plane stress assumes that one direction's stress components are insignificant in comparison to the other two, whereas plane strain assumes that one direction's strain components are insignificant in comparison to the other two. These assumptions limit the analysis to situations in which the distribution of stress or strain remains constant along a particular plane, allowing for simpler mathematical formulations. For complicated three-dimensional problems, this restriction might not apply.

5. Versatility and Straight Superposition: The idea of direct superposition is frequently utilized to investigate the reaction of flexible materials to various applied loads. To calculate the total response, it assumes that the response to each load can be summed linearly. However, materials with non-linear behavior or when the applied loads cause significant deformations may challenge this assumption. In addition, large displacements or material nonlinearity may invalidate the linear superposition assumption for structures.

6. Temperature Impacts: The prologue to anxiety in strong mechanics frequently dismisses the impact of temperature on the material's behavior. Even though this simplification may be reasonable in some situations, it can be a significant limitation when dealing with materials that are sensitive to changes in temperature, like polymers or metals with a high coefficient of thermal expansion. Stress, strain, and structural integrity can be predicted incorrectly if temperature effects are ignored.

STRESS ANALYSIS

Stress examination is a basic part of strong mechanics that deals with the investigation of the dispersion and impacts of inward powers inside a strong material. Solids undergo internal stresses and deformations that have an impact on their mechanical behavior when they are subjected to external loads or forces. It is essential to comprehend and analyze these stresses and strains when designing and analyzing mechanical components and structures to guarantee their safe and effective operation.

The internal forces that develop within a solid material as a result of external loads are measured by stress. It is typically denoted by a symbol and is defined as the force per unit area. Stress can be arranged into various sorts, contingent upon the applied burdens and the subsequent disfigurements. Tensile stress, compressive stress, shear stress, and bending stress are the most typical types of stress.

When a solid experiences elongation as a result of being pulled or stretched, it experiences tensile stress. Compressive pressure, then again, emerges when a strong force is exposed to compressive power, bringing about a diminishing of its aspects. Shear pressure happens when equal powers act digressively across the essence of a material, disfiguring it by sliding or shearing. In structures that are subjected to bending moments, bending stress develops, resulting in compression on one side and tension on the other.

Strain is a proportion of the deformity experienced by a strong material in light of applied pressure. The symbol typically denotes it, which is defined as the ratio of the material's change in length (deformation) to its original length. Strain can be classified into various sorts, including pivotal strain, shear strain, and volumetric strain.

Pivotal strain is related to changes in the length of a material because of pliable or compressive burdens. It is defined as the material's change in length divided by its original length. On the other hand, shear strain is related to a material's shape change or distortion caused by shear stresses. It is calculated by dividing the original length by the angular deformation. By dividing the change in volume by the original volume, volumetric strain can be used to describe the volumetric change caused by stresses applied to a material.

Using the concepts of equilibrium and compatibility, it is necessary to examine the stresses and strains in a solid material. Harmony expects that the number of powers following up on a material towards any path is zero, and the number of minutes about any point is likewise zero. Similarity, then again, guarantees that the disfigurements coming about because of the applied anxieties are viable with the calculation and limit states of the construction.

Stress investigation includes the assurance of the pressure dispersion inside a strong material under different stacking conditions. Analytical calculations, numerical approaches, or experimental testing are typically used to accomplish this. To figure out the stresses and strains in simple geometries and loading configurations, analytical methods use mathematical equations and formulas derived from solid mechanics.

For stress analysis of complex structures and components, numerical methods like the finite element method (FEM) are frequently utilized. Because FEM divides the structure into smaller parts, it is possible to approximate the values of stress and strain at discrete points within the structure. Engineers can optimize their designs for strength and performance by accurately analyzing the behavior of structures under various loading conditions.

Physical prototypes or models are used in experimental testing to directly measure the stresses and strains. This can be accomplished by using load cells, strain gauges, or other sensing devices that record the material's response to the loads that are applied. Validating analytical and numerical models and gaining insight into the behavior of materials in real-world conditions can both benefit from experimental data.

Definition of Stress

Stress is a fundamental concept in solid mechanics that is essential for comprehending the behavior of materials under external loads. It's the internal distribution of forces within a solid material that changes when external loads are applied. Engineers and researchers can predict a material's mechanical behavior, assess its structural integrity, and design components that can withstand anticipated loads by analyzing stress.

The sign is commonly used to denote stress, which is defined as the force exerted per unit area. It addresses the power of inner powers inside a material. At the point when an outside load is applied to a strong, it prompts misshapen Ings and inside powers that demonstration every which way and to every extent. Stress gives a quantitative proportion of these inside powers and decides how they are conveyed inside the material.

Depending on the nature of the loads and deformations that are applied, solid materials can experience a variety of stress types. Tensile stress, compressive stress, shear stress, and bending stress are the most frequently encountered forms of stress in solid mechanics.

Elastic pressure, otherwise called typical pressure, happens when a strong material is exposed to a hub force that will in general extend or pull it apart. It is portrayed by powers acting in inverse directions, opposite to the cross-sectional region of the material. Divide the applied force by the material's initial cross-sectional area to get the tensile stress.

In contrast, when a solid material is subjected to an axial force that tends to compress or push it together, compressive stress results. Compressive stress is characterized by opposing forces acting toward the material's center. It is calculated by dividing the applied force by the material's initial cross-sectional area, just like tensile stress.

When parallel forces act tangentially across a material's faces, shear stress results in the material sliding or shearing. Shear pressure is determined by partitioning the applied power by the area over which it is appropriated. Particularly in applications where materials are subjected to cutting, punching, or twisting forces, it plays a significant role in determining their strength and stability.

When tension and compression forces are exerted on opposing sides of the material, bending stress develops in structures subjected to bending moments. It is calculated by dividing the material's section modulus by the bending moment. Twisting pressure is fundamental to the underlying examination and plan, as it influences the primary steadiness and burden conveying limit of shafts, sections, and different parts.

To break down pressure in strong mechanics, understanding the connection between anxiety and stress is fundamental. Strain alludes to the deformity experienced by the material because of applied pressure. It depicts the material's relative change in size, shape, or volume. The mechanical properties of the material, such as its elasticity, plasticity, and strength, describe the relationship that exists between stress and strain.

The pressure-strain relationship is regularly described by a material's pressure-strain bend, which outlines how the material responds to applied loads. The material deforms in proportion to the applied stress in the elastic region and returns to its original shape when the load is removed. Around here, anxiety is directly related, and the material displays a versatile way of behaving.

Plastic deformation occurs when the applied stress exceeds the material's elastic limit. The plastic deformity is irreversible and prompts an extremely durable change in the shape or size of the material. The pressure strain bend shows a nonlinear conductivity in the plastic locale, with expanding pressure expected to prompt further twisting.

Past the plastic area, the material might arrive at its definitive strength, where cracks or disappointments happen. The pressure strain bend then shows an unexpected drop as the material flops under the applied burden. The maximum amount of stress a material can withstand before breaking is represented by its ultimate strength.

Analytical calculations, numerical approaches, or experimental testing are all ways to conduct stress analysis in solid mechanics. To determine stress distributions in straightforward geometries and loading conditions, analytical methods employ mathematical equations and formulas derived from solid mechanics. These strategies give knowledge into the way, or experimental testing is all ways to conduct stress analysis in solid mechanics. To determine stress distributions in straightforward geometries and loading conditions, analytical methods employ mathematical equations and formulas derived from solid mechanics. These strategies give knowledge into the way materials behave, yet they might be restricted in dissecting complex designs.

For stress analysis of complex structures and components, numerical methods like the finite element method (FEM) are frequently utilized. Because FEM divides the structure into smaller parts, it is possible to approximate the values of stress and strain at discrete points within the structure. Engineers can optimize their designs for strength and performance by accurately analyzing the behavior of structures under various loading conditions.

Physical prototypes or models are used in experimental testing to directly measure the stresses. This can be accomplished by using load cells, strain gauges, or other sensing devices that record the material's response to the loads that are applied. Validating analytical and numerical models and gaining insight into the behavior of materials in real-world conditions can both benefit from experimental data.

In conclusion, stress analysis is an essential component of solid mechanics that provides useful information regarding the internal forces and deformations that occur within materials when subjected to external loads. Engineers and researchers can design structures and components that can withstand anticipated loads and guarantee the safety and dependability of various mechanical systems by gaining an understanding of stress.

Various Types of Stress

Stress is a fundamental idea in mechanics that describes the internal forces that material experiences when it is subjected to external loads or forces. Normal stress, shear stress, and principal stress are all examples of stress manifestations. Understanding these various sorts of pressure is vital for dissecting the mechanical behavior of materials and designs. We will examine the characteristics and effects of each type of stress in this section.

1. Normal Anxiety: The stress that acts perpendicular to a specific cross-sectional area of a material is called normal stress, or axial stress. It occurs when the material experiences stretching or compression due to the application of forces. Typical pressure can be further arranged into tractable pressure and compressive pressure.

When a material is stretched or elongated by an external force, tensile stress develops. This sort of pressure brings about a positive typical pressure value, demonstrating that the material encounters strain. The cause of elongation or deformation in the direction of the applied force is referred to as tensile stress, which is typically indicated by a positive sign.

Compressive pressure, then again, happens when a material is exposed to an outside force that will, in general, pack or abbreviate it. Compressive pressure yields negative qualities, demonstrating that the material is under pressure. The contraction or deformation in the opposite direction of the applied force is caused by compressive stress.

Typical pressure is commonly addressed by the image and is estimated in units of power per unit region, like N/m^2 or Pascal . Because the material's ability to withstand tension or compression is dependent on its strength and ductility, it plays a significant role in determining the structural integrity of materials.

2. Shear Pressure: When forces act tangentially in a parallel direction across the faces of a material, shear stress develops. Sliding or shearing, it deforms the material. Shear stress, in contrast to normal stress, acts parallel to the area. Normal stress is perpendicular to the cross-sectional area. Shear pressure can be imagined as the power per unit area that makes nearby layers of a material slide or distort relative to one another. Shear stress plays a crucial role in determining the material's resistance to deformation and its capacity to withstand shear forces. It is also measured in units of force per unit area, such as N/m2 or Pa.

Analysis of the stability and strength of structures that are subjected to forces that cause shearing, such as bolts, rivets, and welds, requires taking into account shear stress. Understanding shear pressure is especially significant in circumstances where the material's disappointment or misshapen Ing is brought about by the sliding or shearing of its layers.

3. Primary Stress: Chief pressure alludes to the most extreme and least positive aspects of stress experienced by a material at a particular point or area. The evaluation of materials' structural integrity relies heavily on these values, which represent the most extreme stress states. In a three-layered pressure examination, there are for the most part three chief burdens: 1, 2, and 3. 1 addresses the most extreme ordinary pressure, 2 addresses the transitional pressure, and 3 addresses the base typical pressure. The equilibrium and compatibility equations for a given material and loading condition are solved to determine the principal stresses.

Chiefburdensassume a crucialrole in breaking down the disappointment and deformity of materials. The most extreme chief pressure is frequently used to determine if a material will flop under static or cyclic stacking conditions. Engineers can evaluate structures' safety and dependability by comparing the principal stresses to the material's strength.

Significantly, chief burdens can happen under various stacking conditions, like pressure, shear, or a blend of these. They bring significant experience to the basic districts of a construction project and help in upgrading plans for further development of execution and strength.

All in all, stress examination includes the thought of different kinds of pressure, including typical pressure, shear pressure, and chief pressure. Normal stress can be either tensile or compressive and occurs perpendicular to a material's cross-sectional area. Sliding or shearing deformation is the result of shear stress acting parallel to the area. Principal stress is a crucial factor in determining a material's structural integrity because it represents the most extreme states of stress. To design and analyze structures in a way that ensures their safe and effective operation, it is essential to comprehend these various types of stress.

Stress Transformation

Solid Mechanics: An Introduction to Stress and Strain The study of how solid materials behave when subjected to external forces is the focus of the mechanics' subfield known as stress transformation solid mechanics. Stress and strain are fundamental concepts in comprehending the mechanical response of materials and are the subject of this research. Stress alludes to the inside opposition of a material to the outside forces following up on it, while strain estimates the twisting or change in the state of the material coming about because of the applied pressure.

Anxiety examination plays a vital role in different design disciplines, including common, mechanical, and advanced plane design. Specialists and researchers depend on anxiety estimations to configure structures, anticipate their presentation, and guarantee their wellbeing under various stacking conditions. It is essential to take into account stress transformation, which enables the evaluation of stresses in various directions, to analyze and comprehend the behavior of materials.

Stress change includes the transformation of stresses, starting with one direction and then onto the next. Materials' varying strengths and weaknesses in various directions necessitate this transformation, and the applied stresses may act in any direction. Engineers can figure out the maximum and minimum stresses, as well as the orientations of the principal stresses, which are crucial for determining a material's potential for failure, by transforming stresses.

To comprehend pressure change, presenting the idea of stress components is fundamental. There are three components to stress: shear stresses as well as normal stresses. Ordinary burdens act oppositely to a given plane, while shear stresses act digressively to the plane. The ordinary anxieties are additionally ordered as expected pressure in the x-heading (x), typical pressure in the y-course (y), and ordinary pressure in the z-bearing (z). In a similar vein, the shear stresses are denoted by the letters by, is, and six on the by, is, and six planes, respectively.

Two fundamental tenets underpin stress transformation: strain-displacement relationships and equilibrium equations. According to the equilibrium equations, the sum of moments around any point and the sum of forces acting in any direction must both be zero. The material's deformation in terms of its displacement gradient is described by the strain-displacement relationships. The foundation for stress transformation analysis is built on these tenets.

The Mohr's circle is a common stress transformation technique. Engineers can visualize stress transformation and identify the principal stresses with the help of Mohr's circle, a graphical representation. Plotting the normal stresses on the horizontal and shear stresses on the vertical axes is what you need to do. By developing a circle with the middle situated at the normal of the ordinary burdens and a sweep equivalent to a portion of the contrast between the most extreme and least typical burdens, specialists can determine the chief anxieties and their directions.

The direct application of equations is yet another approach to stress transformation. The change conditions relate the anxieties in a single-direction framework to the burdens in another-direction framework. These conditions include geometrical capabilities and permit designers to straightforwardly work out the changed anxieties. Engineers can determine the orientations of the maximum and minimum transformed stresses by utilizing the transformation equations.

The study of materials' behavior under a variety of loading conditions necessitates the use of stress transformation. It enables engineers to design structures that can withstand the applied forces, identify critical locations, and evaluate the state of stress in a material. Engineers can improve the design, reduce the likelihood of failure, and guarantee the structural integrity of various engineering components by comprehending stress transformation.

All in all, stress change is a key idea in strong mechanics that empowers designers to break down and comprehend the way materials behave under various stacking conditions. By switching stresses, starting with one direction framework over completely and then onto the next, designers can decide the most extreme and least burdensome directions. This data is basic for evaluating the disappointment capability of material and planning protected and productive designs. For stress transformation analysis, direct equation application and Mohr's circle are frequently used.

Equations And Methods Used For Stress Transformation, Including Mohr's Circle

Stress change includes the transformation of stresses starting with one direction framework and then onto the next, permitting designers to decide the chief burdens and their directions. For stress transformation, several equations and approaches are utilized, one of which is the well-known graphical approach known as Mohr's circle. Let's get into the specifics of these equations and approaches:

1. Change Conditions: The stresses in one coordinate system are linked to the stresses in another by the transformation equations. The principles of equilibrium and strain-displacement relationships are used to derive these equations. The most commonly utilized change conditions. By knowing the initial stresses and the rotational angle, engineers can directly calculate the transformed stresses using these equations.

2. Mohr's Circle: Mohr's circle is a popular graphic method for transforming stress. Engineers can use it to identify the primary stresses and their orientations, and it provides a visual representation of stress transformation. The following are the steps needed to build Mohr's circle:

- **Step 1:** Draw a graph showing the normal and shear stresses. The horizontal axis is used to plot the normal stresses, while the vertical axis is used to plot the shear stresses.
- **Step 2:** The average of the normal stresses can be found in the center of the circle. Its x-coordinate is the normal of the typical burdens, and its y-coordinate is zero.
- **Step 3:** Decide the span of the circle, which is a portion of the distinction between the greatest and least ordinary burdens.
- **Step 4:** Draw a circle with the center and radius you've determined in the steps before.
- **Step 5:** The primary stresses are represented by the points where the circle intersects the horizontal axis. The greatest chief pressure (1) compares to the furthest right convergence point, while the base chief pressure (2) relates to the furthest left convergence point. The point between the x-pivot and the line connecting the middle to any convergence point addresses the direction of the chief pressure.

Mohr's circle gives a reasonable representation of stress change and improves on the assurance of chief burdens, making it a significant device in designing examinations.

3. Numerical Approaches: For stress transformation, numerical techniques like matrix algebra and tensor notation can be utilized in addition to Mohr's circle. These strategies include controlling pressure parts and utilizing numerical tasks to change burdens. When dealing with complicated loading conditions or when automated calculations are required, these methods are especially helpful.

Stress transformations involving multiple coordinate systems are made simpler by matrix notation, which makes it possible to express the transformation equations succinctly in terms of matrix multiplication.

On the other hand, tensor notation provides a robust mathematical framework for representing and modifying three-dimensional stress and strain components. It empowers specialists to communicate pressure change conditions in a reduced and exquisite structure, working with estimations and examination.

Analyzing and comprehending the behavior of materials under various loading conditions requires these stress transformation equations and methods. They empower architects to decide the most extreme and least extreme anxieties and their directions, which are vital for evaluating primary trustworthiness, planning safe designs, and anticipating disappointment potential.

Strain Analysis

Anxiety examination is a principal idea in the field of strong mechanics, which manages the investigation of the way materials behave when affected by outside powers. Engineers and scientists in a variety of fields, including civil engineering, mechanical engineering, materials science, and structural analysis, need to be aware of stress and strain.

A material's internal resistance to an external force is referred to as stress, and the material's deformation as a result of the stress is referred to as strain. Engineers can predict the mechanical behavior of materials and structures, evaluate their performance, and design effective and dependable systems by analyzing stress and strain.

The material's mechanical properties, such as elasticity, plasticity, and strength, determine the relationship between stress and strain. These properties determine how a material responds to the applied forces and how it disfigures under various stacking conditions. Engineers can calculate and measure these responses with the help of stress and strain analysis, which provides useful data for design and analysis.

Analysis of Stress

Stress investigation includes determining the inside powers and stresses inside a material or construction. These inner powers can result from outer burdens, temperature changes, or other ecological elements. The pressure experienced by a material is commonly addressed as a power for every unit region and is estimated in units of tension, like pascals or pounds per square inch (psi).

Stress can be arranged into various sorts in light of the idea of the applied powers and the subsequent deformity. The three fundamental sorts of pressure are:

- **Pliable Pressure:** Elastic pressure happens when a material is exposed to forces that attempt to prolong or extend it. Materials under tension, such as a rope being pulled from both ends, typically exhibit this kind of stress. Malleable pressure is communicated as certain qualities.
- **Compressive Pressure:** Compressive pressure is something contrary to tractable pressure. It emerges when a material is exposed to powers that, more often than not, pack or abbreviate it. This sort of pressure is experienced, for instance, when a section upholds a weighty burden. Compressive pressure is additionally communicated as certain qualities.

- **Shear Pressure:** Shear pressure is created when two powers act in line with one another yet in inverse bearings, making the material distort by sliding or shearing. Material deformation and failure are caused by shear stress in planes parallel to the applied forces. The Greek letter tau is commonly used to represent shear stress, which is also measured in pressure units.

Strain Examination

Quantifying a material's deformation under applied stress is the primary goal of strain analysis. The ratio of the material's change in length—also known as deformation—to its original length is what is referred to as strain. Macrostrains and parts per million (ppm) are common ways to represent strain, which is a dimensionless quantity.

Strain, like stress, can be broken down into the following categories:

- **Tension Damage:** Pliable strain alludes to the lengthening or expansion of material along the course of applied ductile pressure. Positive values serve as a symbol for it.
- **Compressive Strain:** When a material shortens or contracts in the direction of applied compressive stress, this is called compressive strain. Positive values also represent compressive strain.
- **Shear Strain:** A material's change in shape or distortion when subjected to shear stress is referred to as a shear strain. It is typically measured in radians or degrees and is characterized by angular deformation.

Relationship Between Stress and Workout

A material's mechanical properties are what link the stress and strain it experiences. A stress-strain curve, which shows how the material responds to increasing stress, typically describes this relationship.

Hooke's law states that the material has a linear relationship between stress and strain in the elastic portion of the stress-strain curve. The material will return to its original shape when the applied forces are removed, as evidenced by this linear relationship, indicating that it is exhibiting elastic behavior.

Plastic deformation may occur when the stress exceeds a certain point, known as the yield strength. Around here, the pressure strain bend becomes nonlinear, and the material encounters long-lasting misshapen Ing even after the powers are taken out.

Finally, the material may reach its ultimate strength and fracture or fail if the applied stress continues to rise. The strength, stiffness, ductility, and other mechanical properties of the material can all be learned from the stress-strain curve.

Utilisations of the Anxiety Examination

In the fields of engineering and material science, stress and strain analysis can be used in a variety of ways. A few key applications include:

- **Analyses of the Structure:** Anxiety examination is significant for surveying the exhibition and wellbeing of designs, like structures, extensions, and airplanes. To guarantee that the structures can withstand the anticipated loads and environmental conditions, engineers examine the distribution of stress and strain.
- **Material Selection and Design:** Understanding the anxiety conductivity of various materials permits specialists to choose suitable materials for explicit applications. They can assess the mechanical properties and execution of materials to guarantee their ideal plan and solidity.
- **Analyses of Failures:** Anxiety examination assumes a fundamental role in disappointment examination and counteraction. By examining pressure fixations, material shortcomings, and the variables prompting disappointment, specialists can foster systems to keep away from comparative disappointments later on.
- **Producing Cycles:** In manufacturing processes, stress, and strain analysis are used to make sure that materials are formed, shaped, and assembled correctly. Manufacturers can improve product quality and optimize production methods by anticipating and controlling the distribution of stress and strain.

Definition of Strain

In strong mechanics, strain refers to the measure of deformation experienced by a material in response to applied stress. It quantifies the change in shape or size of a material due to external forces following up on it. Strain is a dimensionless amount and is commonly expressed as a proportion or a percentage.

The relationship between strain and stress is governed by the material's mechanical properties, like its elasticity and stiffness. When stress is applied to a material, it induces deformation, resulting in strain. This relationship is described by Hooke's regulation, which states that, as far as possible, the strain produced in a material corresponds directly to the applied stress.

Mathematically, Hooke's regulation is expressed as:

$$\sigma = E\varepsilon$$

Where:

σ = represents stress (force per unit area),

E is the modulus of elasticity (otherwise called Young's modulus),

Also, ε denotes strain.

As per Hooke's regulation, the strain is directly relative to the stress, and the steady state of proportionality is the modulus of elasticity. The modulus of elasticity characterizes the material's resistance to deformation and determines how it responds to stress. Different materials have different modulus of elasticity values, which reflect their stiffness and capacity to return to their unique shape after deformation.

It is vital to note that Hooke's regulation turns out as expected inside the elastic's furthest reaches. Once the applied stress exceeds it as far as possible, the material might undergo plastic deformation, where strain becomes permanent even after the removal of stress. In the plastic deformation range, the relationship between stress and strain is nonlinear, and different material models are used to describe the behavior.

Overall, the relationship between strain and stress in strong mechanics is fundamental for understanding the deformation and mechanical response of materials under external forces. By dissecting this relationship, engineers can predict material behavior, assess primary integrity, and design components that meet desired performance criteria.

Particular strains are frequently encountered in solid mechanics, in addition to the general classifications of tensile, compressive, and shear strains. Axial strain, shear strain, and volumetric strain are examples of these. Let's look at each of these categories:

Pivotal Strain

The deformation that takes place in the direction of the applied stress or along an object's axis is referred to as axial strain, which can also be referred to as linear strain or longitudinal strain. It talks about a material's length changing from its initial length. Pivotal strain can be either malleable or compressive, contingent upon the applied powers. Elastic Pivotal Strain: Pliable pivotal strain happens when a material stretches along the hub affected by ductile pressure. It is calculated by dividing the change in length by the original length, and its values are positive.

Axial Compression Strain: When a material experiences compressive stress, it experiences compressive axial strain when it contracts or shortens along the axis. Positive values also represent it because they indicate a shorter length.

Shear Strain

When adjacent layers of a material slide or deform concerning one another, shear strain, also known as tangential strain or angular strain, results. It symbolizes the material's change in shape or distortion. The angular deformation that characterizes shear strain is measured in degrees or radians. When forces act in opposite directions while being parallel to one another, shear stress results in shear strain. The shear strain is calculated by dividing the original angle by the change in angle between the deformed and undeformed directions.

Volumetric Strain

The change in a material's volume caused by stress is referred to as volumetric strain, as the name suggests. It represents the constriction or development of the material this way and that. Volumetric strain is impacted by the three chief strains: pivotal strain, sidelong strain, and longitudinal strain.

Positive or negative volumetric strain can occur. While negative volumetric strain indicates a decrease in volume, positive volumetric strain indicates an increase in volume. It is determined by adding up the three primary strains.

The Poisson's ratio, which describes the ratio of lateral strain to axial strain in a material subjected to axial stress, is related to the relationship between axial strain, lateral strain, and longitudinal strain. The material's response to stress and strain is significantly influenced by Poisson's ratio.

To analyze and predict the behavior of materials under various loading conditions, it is essential to comprehend the various strain types—axial strain, shear strain, and volumetric strain, for instance. These strain gauges are utilized by scientists and engineers to design structures, evaluate the performance of materials, and guarantee the safety and dependability of engineering systems.

HOOKE'S LAW AND ELASTICITY

In the discipline of solid mechanics, which studies how solid materials behave under a variety of mechanical loads, stress and strain are fundamental concepts. When designing and analyzing structures to ensure their safety and performance, it is essential to comprehend stress and strain. One of the key rules that oversee the way materials behave under load is Hooke's Regulation, which portrays the connection between anxiety and flexible materials. The ability of a material to deform under stress and return to its original shape when the load is removed is studied using this law as the foundation.

Hooke's Regulation

Hooke's Regulation, named after the prestigious physicist Robert Hooke, expresses that the pressure in a strong material is straightforwardly corresponding to the strain prompted in the material, as long as the material remains parts inside its versatile cut-off. Hooke's Law can be expressed mathematically as:

$$\sigma = E\varepsilon,$$

where denotes strain, the elastic modulus (also referred to as Young's modulus), and stress. That's what this straight relationship suggests: assuming the pressure applied to a material pair, the subsequent strain will likewise be twofold. Similarly, the strain will decrease proportionally if stress is decreased by half.

Flexibility and Versatile Materials

Flexibility is the property of a material that empowers it to distort under pressure and, in this way, return to its unique shape when the pressure is removed. Elastic materials are materials that behave in this way. Examples of ordinarily experienced flexible materials incorporate metals like steel and aluminum, as well as specific polymers and elastomers.

Deformation occurs when an elastic material is subjected to stress. This deformation can be compressive (which makes the material shorter) or tensile (which makes the material longer). As long as the elastic limit is not exceeded, Hooke's Law states that the strain caused by this deformation is directly proportional to the stress applied.

Material Properties and Elastic Modulus The material's stiffness or resistance to deformation is characterized by its elastic modulus, or Young's modulus (E). It shows how much stress is needed to cause a certain strain on a material. Each material's Young's modulus is measured in pressure units, such as pascals (Pa) or gigapascals (GPa), typically.

The varying resistance to deformation of various materials is reflected in their various elastic moduli. Steel, for example, has a high Young's modulus, which means that it is stiff and needs a lot of stress to cause a certain strain. Rubber, on the other hand, is much more flexible and easier to deform due to its low Young's modulus.

The versatile modulus likewise gives significant data about the material's reaction to pressure. Ceramics, for instance, are brittle materials with a relatively high modulus that easily fracture under stress beyond their elastic limit. However, ductile materials, such as many metals, exhibit plastic deformation before failure and have lower moduli.

Elastic Limits and Hookean Behaviour Even though Hooke's Law holds that there is a linear relationship between stress and strain, the elastic portion of a material's stress-strain curve is the only area where this behavior can occur. Beyond which it exhibits non-linear behavior and does not return to its original shape upon unloading, every material has an elastic limit. When as far as possible is outperformed, super-durable twisting or pliancy happens.

Engineering design must take the elastic limit into account because structures and parts shouldn't be stressed beyond this limit. Working a material past its flexible cutoff can prompt primary disappointment, decreased execution, and compromised security.

The elastic modulus, otherwise called Young's modulus (E), is a material property that quantifies the stiffness or inflexibility of a strong material. It describes the relationship between stress and strain inside the elastic region of a material's stress-strain curve.

Mouthful's modulus is defined as the proportion of stress to strain and is mathematically represented as:

$$E = \sigma/\varepsilon$$

where E is the elastic modulus, is the stress applied to the material, and is the resulting strain. The elastic modulus is normally measured in units of pressure, like pascals or gigapascals (GPa).

The elastic modulus reflects a material's capacity to resist deformation under applied stress. Materials with a high elastic modulus are considered firm or inflexible, requiring huge stress to produce a given strain. Then again, materials with a low elastic modulus are more flexible and deform more easily under the same stress.

Different materials exhibit a wide range of elastic moduli due to variations in their nuclear or molecular structure, holding types, and overall material composition. For example, metals like steel and aluminum have relatively high elastic moduli, showing their stiffness and resistance to deformation. These materials are often used in underlying applications that require strength and a burden-bearing limit.

Interestingly, materials like rubber or elastomers have low elastic moduli, making them profoundly flexible and capable of critical deformation under stress. These materials are generally used in applications that require elasticity, like rubber groups or gaskets.

The elastic modulus likewise provides valuable data about a material's response to stress. It helps determine how much a material will deform when subjected to an applied load and how it will recover its unique shape once the heap is removed. Materials with higher elastic moduli tend to exhibit more elastic behavior, returning to their unique shape with negligible permanent deformation. Interestingly, materials with lower elastic moduli might undergo more plastic or permanent deformation under stress.

The elastic modulus is a significant parameter in engineering design and material selection. It helps engineers predict and analyze the behavior of structures and components under different stacking conditions. By knowing the elastic modulus of a material, engineers can determine its reasonableness for specific applications and ensure that the material can endure the expected stresses without excessive deformation or failure.

All in all, the elastic modulus is a fundamental material property that characterizes a material's stiffness or unbending nature. It describes the relationship between stress and strain inside the elastic region and provides insight into a material's deformation behavior. Understanding the elastic modulus helps engineers design safe and efficient structures by selecting materials with appropriate stiffness for specific applications.

Hooke's Regulation is of huge importance in relating stress and strain in linear elastic materials. It provides a fundamental understanding of the behavior of materials under mechanical burdens inside their elastic breaking point. Here are some key focuses that feature the significance of Hooke's Regulation:

- **Linear Relationship:** Hooke's Regulation states that stress is directly correlated to strain inside the elastic furthest reaches of a material. This linear relationship considers a direct and predictable examination of the material's response to applied loads. Engineers and scientists can use this relationship to determine the magnitude of stress and strain in a material under different stacking conditions.
- **Predictability:** The linear relationship described by Hooke's Regulation enables engineers to predict the behavior of materials under different stress levels. By knowing the elastic modulus (Mouthful's modulus) of a material, the stress-strain relationship can be used to calculate the resulting strain for a given stress. This predictability is imperative for designing structures and components, ensuring that they will perform as intended and remain inside their elastic cutoff points.
- **Elastic Cutoff Determination:** Hooke's Regulation helps in determining the elastic furthest reaches of a material, which is the place where it changes from elastic deformation to plastic deformation. By subjecting a material to steadily increasing stresses and checking the resulting strains, as far as possible, the place where the linear relationship described by Hooke's Regulation is not substantial as of now can be identified. This data is significant for understanding the material's limits and designing structures that don't exceed their elastic cutoff points.
- **Material Characterization:** Hooke's Regulation provides a means

to characterize different materials based on their elastic behavior. The elastic modulus (Mouthful's modulus) is a material property derived from Hooke's Regulation, and it quantifies a material's stiffness or resistance to deformation. By looking at the elastic moduli of different materials, engineers can select appropriate materials for specific applications based on their desired stiffness or flexibility.

- **Underlying Investigation:** Hooke's Regulation structures the establishment for examining the primary behavior of elastic materials. By applying Hooke's Regulation, engineers can determine the stress conveyance inside a structure under load and assess its primary integrity. This examination helps in designing structures that can endure expected loads without excessive deformation or failure.
- **Material Testing:** Hooke's Regulation guides the experimental testing of materials to determine their mechanical properties. By subjecting a material to controlled loads and measuring the resulting strains, engineers can validate Hooke's Regulation and get material-specific parameters like the elastic modulus. This data is indispensable for accurate material characterization and design computations.

In outline, Hooke's Regulation assumes a basic role in relating stress and strain in linear elastic materials. It provides a fundamental understanding of material behavior, helps with design and examination, facilitates material characterization, and guides material testing. By using Hooke's Regulation, engineers and scientists can effectively predict and control the deformation of materials under mechanical burdens, ensuring the safe and efficient design of structures and components.

Experimental Techniques for Stress and Strain Analysis

Stress and strain examination plays a vital role in understanding the mechanical behavior of materials and structures. Experimental techniques offer valuable insights into the dispersion of stresses and strains under different stacking conditions. This chapter provides a comprehensive overview of different experimental techniques used for stress and strain investigation, featuring their principles, advantages, and impediments. By exploring these techniques, researchers, engineers, and scientists can acquire a deeper understanding of material behavior and make informed decisions for design and enhancement purposes.

- **Strain Gauges:** Strain gauges are widely used in experimental stress examinations due to their versatility and exactness. These devices measure strain by exploiting the change in electrical resistance of a wire or foil when subjected to deformation. They can be bonded to the surface of a test specimen and provide direct strain measurements. Strain gauges are available in different structures, like wire, foil, and semiconductors. They offer excellent sensitivity and are suitable for measuring both static and dynamic strains. However, the holding process can be time-consuming and requires expertise. Furthermore, temperature effects and mechanical hysteresis can introduce errors in the measurements.

- **Photoelasticity:** Photoelasticity is an optical technique used to visualize stress appropriation in transparent materials. It relies on the principle of birefringence, where the refractive index of a material changes under stress. By presenting polarised light and examining the resulting interference patterns, engineers can determine stress concentrations and the magnitude of the chief stresses. Photoelasticity is advantageous as it provides a full-field stress investigation and can be used to concentrate on complex stress states. However, it is limited to transparent materials, and the interpretation of fringe patterns requires expertise.
- **Moiré Interferometry:** Moiré interferometry is an optical technique that utilizes the interference patterns produced by superimposing two comparable matrices with a slight misalignment. It is used to measure displacement and strain fields with high sensitivity. Moiré interferometry can be applied to both 2D and 3D stress investigations and offers full-field measurements. It is non-contact and suitable for real-time checking. However, complex setups, susceptibility to environmental disturbances, and information interpretation challenges pose certain constraints.
- **Advanced Image Correlation:** Computerized Image Correlation (DIC) is a widely used technique that tracks the displacement of features on the surface of a test specimen using image processing calculations. By dissecting the changes in pixel intensities, DIC provides full-field strain measurements. A non-contact technique offers high resolution and subpixel precision. DIC can be applied to a wide range of materials and stacking conditions, including large deformations. However, it requires top-notch images, and the presence of surface irregularities or impediments can affect precision.
- **Acoustic Emission:** Acoustic emission (AE) is a technique that involves observing the ultrasonic waves generated by a material under stress. These waves convey data about break spread, plastic deformation, and other material changes. AE can be used for both qualitative and quantitative examination, giving insight into the

condition and integrity of a structure. It is a non-destructive testing method that can be applied during the entire life cycle of a component. However, the interpretation of AE signals requires expertise and environmental noise can interfere with measurements.

Strain Measurement Methods

Strain measurement assumes an essential role in different fields, including engineering, materials science, and biomechanics. It involves evaluating the deformation experienced by an object under external burdens or internal stresses. Accurate strain measurement is essential for designing and upgrading structures, predicting failure, and evaluating material properties. Over the years, numerous strain measurement methods have been developed, each with its advantages and impediments. In this chapter, we provide a comprehensive overview of normally used strain measurement methods, examining their principles, applications, and considerations.

Electrical Resistance Strain Gauges

Electrical resistance strain gauges are one of the most widely used and versatile strain measurement methods. They operate on the principle that the electrical resistance of a wire changes as it undergoes deformation. Strain gauges consist of a fine wire or foil network pattern attached to a flexible sponsorship material. When the strain gauge is bonded to the surface of a test specimen, any deformation experienced by the specimen causes a change in resistance, which can be measured using a Wheatstone bridge circuit. Electrical resistance strain gauges offer high sensitivity, a wide strain range, and excellent frequency response, making them suitable for both static and dynamic strain measurements.

- **Optical Methods:** Optical methods provide non-contact and full-field strain measurement capabilities, making them valuable for applications where the direct attachment of strain gauges is illogical or undesirable. Moiré interferometry, computerized image correlation, and holographic interferometry are some of the optical techniques used for strain measurement. Moiré interferometry relies on the interference pattern formed between two superimposed frameworks to measure strain. Advanced image correlation compares images of an object captured before and after deformation to calculate strain values. Holographic interferometry employs laser light to generate interference fringes, enabling precise strain investigation. Optical methods offer high precision, high resolution, and the capacity to capture transient events.

- **Acoustic Methods:** Acoustic methods utilize sound waves to measure strain in materials. These methods can be classified into two categories: ultrasonic and acoustic emission. Ultrasonic strain measurement involves communicating ultrasonic waves through a material and breaking down the changes in wave generation caused by strain. It is especially useful for measuring strain in inaccessible or unsafe environments. Acoustic emission, then again, detects high-frequency elastic waves emitted during the deformation of a material. It is effective in observing break commencement and spread, making it suitable for primary health-checking applications.
- **Mechanical Extensometers:** Mechanical extensometers directly measure strain by observing the change in length or deformation of a specimen. These devices commonly consist of two arms with movable ends that attach to the specimen's surface. As the specimen deforms, the arms move, and the strain is determined based on the change in a manageable distance. Mechanical extensometers offer straightforwardness, strength, and high precision. They are generally used in tensile testing machines and for measuring strains in structures under static or semi-static loads.
- **Piezoelectric Transducers:** Piezoelectric transducers convert mechanical strain into an electrical signal, considering strain measurement in unique applications. When subjected to strain, piezoelectric materials generate an electric charge corresponding to the applied strain. This charge can be measured using charge amplifiers or charge converters. Piezoelectric transducers have a high-frequency response, a wide unique range, and excellent sensitivity. They find applications in underlying health checking, vibration examination, and effect testing.

Stress Measurement Methods

In the present high-speed and demanding world, stress has become an inevitable part of our lives. The capacity to accurately measure and assess stress levels is vital for understanding its effect on people and implementing effective stress management techniques. This chapter provides a comprehensive overview and examination of different stress measurement methods normally used in research and clinical settings.

- **Self-Report Measures:** Self-report measures are widely employed in stress research due to their ease of organization and cost-effectiveness. These measures rely on people's subjective assessments of their stress levels and regularly involve questionnaires or scales. The Perceived Stress Scale (PSS) and the Stressful Life Events Scale (SLES) are two well-known self-report measures. Albeit self-report measures provide valuable bits of knowledge into a person's perceived stress, they might be influenced by elements, for example, response predisposition and individual differences in self-awareness.

- **Physiological Measures:** Physiological measures offer an objective way to deal with stress assessment by examining the physiological responses associated with stress. These measures assess changes in different physiological parameters, including heart rate, circulatory strain, cortisol levels, and skin conductance. A heart rate changeability (HRV) examination, for example, examines the fluctuation in time intervals between successive heartbeats and provides bits of knowledge about autonomic nervous system movement. Electrodermal measurement assesses changes in sweat organ action as a mark of sympathetic nervous system excitement. Physiological measures offer valuable real-time information but require specialized equipment and expertise for accurate interpretation.
- **Behavioral Measures:** Behavioral measures center around assessing observable behaviors that might indicate stress. These measures include direct observation, behavioral coding, and examination of specific behaviors associated with stress, for example, fidgeting, nail gnawing, or changes in speech patterns. Behavioral measures can be especially useful in naturalistic settings or when concentrating on stress in specific contexts, like the workplace or social interactions. However, they might be limited by the presence of perplexing elements and observer biases.
- **Biochemical Measures:** Biochemical measures involve the examination of stress-related biomarkers in natural samples, like blood, spit, or urine. These measures provide bits of knowledge about the physiological changes that happen in response to stress. Cortisol, a hormone released by the adrenal organs during stress, is an ordinarily measured biomarker. Other biomarkers, for example, neuropeptides and incendiary markers, can likewise be assessed. Biochemical measures offer a more direct assessment of stress-related physiological processes; however, they require research facility examination and may not reflect acute stress responses.
- **Neuroimaging Measures:** Advancements in neuroimaging techniques have opened up new avenues for concentrating on the neural correlates of stress. Utilitarian magnetic resonance imaging (fMRI), positron emission tomography (PET), and electroencephalography (EEG) are among the regularly used neuroimaging methods. These measures provide insights into

mind-enactment patterns and connectivity associated with stress. Neuroimaging measures offer a deeper understanding of the underlying neural mechanisms of stress but require sophisticated equipment, expertise, and controlled environments.

Methods for measuring stress encompass different techniques, including the use of burden cells, pressure transducers, and strain-gauge-based stress examination. These methods are regularly employed in engineering and material science disciplines to assess the mechanical stress experienced by objects or materials. This section will talk about each of these measurement methods in more detail:

- **Load Cells:** Load cells are devices designed to measure forces or loads applied to them. They are widely used in stress examinations to determine the magnitude of external forces following up on a structure. Load cells regularly comprise a sensing element, often made of metal or semiconductor material, that deforms in response to applied loads. This deformation generates a change in electrical resistance, capacitance, or voltage, which can be measured and calibrated to determine the applied force. Load cells can be categorized into different types based on their design and working principles, for example, strain gauge load cells, water-powered load cells, and pneumatic burden cells. Strain gauge load cells are the most commonly used type and rely on strain gauges (discussed later) to measure the deformation caused by the applied burden. They offer high exactness and are suitable for both static and dynamic burden measurements.
- **Pressure Transducers:** Pressure transducers are devices used to measure liquid or gas pressures. They are frequently utilized to assess stress in liquid systems like pipelines, water-powered systems, or air compressors. Pressure transducers convert the applied pressure into an electrical signal that can be calibrated and quantified. Pressure transducers employ different principles for pressure measurement, including piezoresistive, capacitive, or piezoelectric effects. Piezoresistive pressure transducers use a strain-sensitive resistor, for example, a silicon strain gauge, to measure the little changes in resistance resulting from the applied pressure. Capacitive pressure transducers utilize the changes in capacitance caused by the deflection of a stomach under the applied pressure. Piezoelectric pressure transducers rely on the piezoelectric effect, where a gem generates an electrical charge corresponding to the applied pressure.
- **Strain-Gauge-Based Stress Investigation:** Strain gauges are widely employed in stress examinations to measure the deformation

or strain experienced by an object or material. They consist of a thin wire or foil, regularly made of a metallic material, which undergoes resistance changes when subjected to strain. Strain gauges are attached to the surface of a structure, and as the structure experiences deformation, the strain gauge deforms with it, resulting in a change in electrical resistance. The change in resistance is relative to the strain and can be measured using a Wheatstone bridge circuit or a strain pointer. Strain gauges are profoundly sensitive and can provide detailed data about the dissemination of stress or strain across a structure. They are used in a variety of applications, including underlying examination, load testing, and material testing.

RESULTS

- **Experimental Setup:** To direct the stress and strain examination, a strong specimen of a selected material was subjected to controlled stacking conditions. The material properties, including Mouthful's modulus (E), Poisson's proportion, and yield strength, were determined before the experiments. The specimen was loaded using a universal testing machine, and the resulting stress and strain were measured using strain gauges and burden cells. The experiments were performed under different stacking conditions, including tensile, compressive, and shear stacking, to capture different stress and strain states.
- **Tensile Stacking:** Under tensile stacking, the specimen experienced an increase along the stacking direction (pivotal strain). The stress-strain curve showed a linear elastic region until reaching the yield strength, beyond which the material exhibited plastic deformation. The ultimate tensile strength (UTS) represented the greatest stress the material could endure before failure. The elongation at break provided insight into the flexibility of the material.
- **Compressive Stacking:** On account of compressive stacking, the specimen experienced a decrease along the stacking direction. Like tensile stacking, the material demonstrated linear elastic behavior until reaching the yield strength, followed by plastic deformation. The compressive strength represented the greatest stress the material could endure before failure. The compressive strain indicated the deformation limit and security of the material under compression.
- **Shear Stacking:** Shear stacking caused the material to deform by sliding one layer of iotas or molecules over another. The shear stress-strain relationship followed a linear elastic behavior until reaching

the shear yield strength, after which the material experienced plastic deformation. The shear modulus (G) provided bits of knowledge about the material's resistance to shear deformation.

- **Hooke's Regulation:** The stress-strain relationship in the linear elastic region obeyed Hooke's Regulation, which states that stress is directly related to strain. The slope of the stress-strain curve in the linear region represented the material's Young's modulus (E). A higher Young's modulus indicated a stiffer material with less elastic deformation under applied stress.
- **Failure Investigation:** A failure investigation was performed to identify the mode and mechanism of material failure under different stacking conditions. The stress and strain appropriation near the weak spot provided insights into the material's fracture behavior, like a brittle or ductile fracture. The failure stress and strain values were critical in determining the material's safety component and designing structures with adequate strength.
- **Suggestions and Engineering Applications:** The results of stress and strain investigations have critical ramifications for different engineering applications. By understanding the material's behavior under different stacking scenarios, engineers can design structures that can endure anticipated stresses while limiting the risk of failure. The information obtained from stress and strain examination can be used to optimize the material selection, assess primary integrity, and predict the service life of components subjected to cyclic stacking.

CONCLUSION

All in all, the investigation of stress and strain in strong mechanics assumes a critical role in understanding the behavior of materials under external forces. All through this examination, we have explored the fundamental concepts of stress, strain, and their interrelation, as well as the different theories and models that help in predicting and dissecting the mechanical response of solids.

Stress, defined as the internal resistance experienced by a material due to applied external forces, is a basic parameter that determines the material's capacity to endure deformation. Different types of stress, including typical stress, shear stress, and hub stress, have unmistakable effects on the behavior of materials. Understanding stress conveyance and the variables that influence it is imperative for designing structures that can endure different stacking conditions.

Then again, strain refers to the deformation experienced by a material when subjected to stress. It is a measure of how much a material changes shape or size in response to applied forces. The relationship between stress and strain is described by constitutive equations, which differ for different materials and are often nonlinear. The investigation of strain helps engineers and scientists evaluate the mechanical properties of materials and predict their behavior under different stacking scenarios.

One of the fundamental concepts in strong mechanics is Hooke's regulation, which states that the stress is directly relative to the strain inside the elastic furthest reaches of a material. This linear relationship takes into consideration the determination of material properties, for example, Young's modulus, which quantifies the stiffness of a material. However, once as far as possible is exceeded, the material enters the plastic region, where permanent deformation happens. As of now, the stress-strain relationship is nonlinear and irreversible.

The investigation of stress and strain likewise involves the examination of stress concentrations, which are areas inside a structure where stress levels are fundamentally higher than the average. These stress concentrations can lead to underlying failure if they are not properly accounted for in the design process.

Engineering, as a multidisciplinary field, deals with the design, development, and implementation of different structures and systems. It encompasses a wide range of uses, from common infrastructure and mechanical components to electronic circuits and aerospace vehicles. In chasing safe and efficient designs, engineers rely on stress and strain examination to assess the behavior and integrity of materials and structures under different stacking conditions. This essay features the urgent importance of stress and strain examination in engineering applications, shedding light on its role in ensuring underlying reliability, advancing designs, and preventing horrendous failures.

Understanding stress and strain: Stress and strain are fundamental concepts in engineering mechanics that describe the response of materials to external forces. Stress is defined as the internal resistance experienced by a material due to applied forces, while strain refers to the resulting deformation or change in shape. Stress and strain investigation involves measuring these responses and determining their effects on the performance and sturdiness of engineering structures. By investigating stress and strain conveyances, engineers can acquire valuable bits of knowledge about the behavior of materials, identify potential failure points, and predict their overall performance. This knowledge enables engineers to make informed decisions during the design and development phases, ultimately ensuring the safety and reliability of structures.

Primary Integrity and Safety: Stress and strain examination plays a crucial role in evaluating the underlying integrity and safety of engineering systems. By subjecting structures to different burden scenarios, engineers can determine the stress levels and deformations experienced by different components. This examination considers the identification of basic areas prone to failure, for example, high-stress concentrations, fatigue zones, or regions susceptible to clasping or fracture. Furthermore, stress and strain investigations help engineers assess the appropriateness of materials for specific applications. Different materials exhibit unique stress-strain behaviors, and understanding their cutoff points is urgent for selecting appropriate materials for a given design. By considering variables like yield strength, ultimate tensile strength, and pliability, engineers can ensure that materials can endure the anticipated burdens without compromising their underlying integrity.

Moreover, stress and strain examination support verifying the compliance of structures with relevant design codes and guidelines. These codes often specify the most extreme allowable stresses and deformations for different materials and stacking conditions. By contrasting calculated stresses and strains with these criteria, engineers can ensure that the design meets safety requirements and minimize the risk of failure during the structure's service life.

Streamlining and performance enhancement: Stress and strain examinations are not limited to ensuring safety but also play an essential role in improving engineering designs and enhancing performance. By examining stress and strain distributions, engineers can identify areas of high-stress concentration, which might lead to premature failure. This data guides the refinement of designs, taking into consideration the reduction of stress concentrations through the presentation of extra reinforcements or adjustments to the geometry. Furthermore, stress and strain examinations help with the evaluation and correlation of different design alternatives. By measuring the stresses and deformations for different design choices, engineers can identify the most efficient and financially savvy arrangements. This process helps in reducing material usage, limiting weight, and streamlining underlying performance, leading to improved efficiency and reduced assembling costs.

Stress and strain investigations likewise uphold the prediction of the service life and strength of engineering structures. By subjecting materials to cyclic stacking conditions, for example, in fatigue examination, engineers can assess the aggregation of damage and determine the expected lifespan of the structure. This data is essential for preventive maintenance and helps in staying away from unexpected failures that could result in critical economic and human consequences.

2

Carbon Fiber and Potential Drop Method of Solid Mechanics

ABSTRACT

Carbon fiber-reinforced composites have gained huge attention in the field of strong mechanics due to their exceptional mechanical properties and lightweight characteristics. This chapter provides a comprehensive review of the use of carbon fiber in strong mechanics, zeroing in on the use of the potential drop method for dissecting the mechanical behavior of carbon fiber-reinforced composites. The potential drop method is a non-destructive testing technique that considers the measurement of break commencement and engendering in materials.

This chapter begins with a prologue to carbon fiber and its unique properties, featuring its excellent strength-to-weight proportion, high stiffness, and erosion resistance. It further discusses the assembling techniques employed in carbon fiber creation, including the carbonization and graphitization processes. Also, the chapter provides an overview of different carbon fiber-reinforced composite structures and their applications in industries like aerospace, automotive, and athletic gear.

The potential drop method is then described exhaustively, illustrating its principle, advantages, and constraints. The technique involves running a current through a specimen and measuring the voltage drop across it, considering the detection and characterization of breaks and defects. The chapter explores the different arrangements and setups utilized in potential drop testing, including the four-point method, the two-probe method, and the alternating current potential drop technique.

Furthermore, the chapter discusses the specific challenges and considerations when applying the potential drop method to carbon fiber-reinforced composites. Factors like the electrical conductivity of carbon fibers, the effect of lattice materials, and the influence of fiber orientation on break detection are addressed. It likewise features the importance of proper specimen preparation, electrode placement, and information examination techniques in ensuring accurate and reliable results.

All in all, this chapter provides a comprehensive overview of carbon fiber and its applications in strong mechanics, with a specific spotlight on the potential drop method for break detection and characterization in carbon fiber-reinforced composites. The data presented in this review can serve as a valuable resource for researchers, engineers, and practitioners in the fields of materials science and mechanical engineering.

***Keywords**: carbon fiber, potential drop method, strong mechanics, composite materials, non-destructive testing, break detection, mechanical properties.*

INTRODUCTION

The potential drop method and carbon fiber are two important aspects of solid mechanics. Numerous industries, including sports, the automotive industry, and aerospace, have been transformed by carbon fiber, which is renowned for its exceptional mechanical properties and light weight. The potential drop method, on the other hand, is a non-destructive testing method for determining the structural integrity of materials, such as carbon fiber composites. Researchers, engineers, and designers in the field of solid mechanics must have a thorough understanding of carbon fiber's properties, applications, and potential drop methods. Carbon fiber is a composite material made out of carbon strands implanted in a lattice material, ordinarily a polymer. The carbon strands are incredibly dainty, with distances across going from 5 to 10 micrometers. These filaments are made out of lengthy, dainty strands of carbon molecules, firmly reinforced together in a glasslike structure. Carbon fiber is one of the strongest and stiffest materials available because of its exceptional strength-to-weight ratio due to the unique arrangement of carbon atoms. The mechanical properties of carbon fiber are surprising. Because of its high tensile strength, it can withstand pulling forces without bending or breaking. Additionally, carbon fiber has a high modulus of elasticity, which indicates that it is extremely rigid and can withstand substantial bending forces without deforming. Additionally, carbon fiber is suitable for applications involving cyclic loading due to its excellent fatigue resistance. The uncommon mechanical properties of carbon fiber have prompted a large number of uses. Carbon fiber composites are frequently used in the production of aircraft parts like wings, fuselages, and tail sections in the aerospace industry. Carbon fiber's light weight contributes to the aircraft's overall weight reduction, which improves fuel economy and increases payload capacity. Additionally, the high strength of carbon fiber permits enhanced structural integrity, ensuring flight safety. The car business has additionally embraced carbon fiber composites because of their beneficial properties. Automobile manufacturers can reduce weight while maintaining structural strength by incorporating carbon fiber into the vehicle's structures. As a result of this weight reduction, acceleration can be accelerated more quickly, and the vehicle can handle better. Also, carbon fiber is utilized in the development of sports vehicles and elite execution vehicles to achieve the most extreme speed and dexterity. Carbon fiber's advantages have also been recognized by sports equipment manufacturers. Carbon fiber components are now found in bicycles, tennis rackets, golf clubs, and other sporting goods to improve performance. The lightweight idea of carbon fiber permits competitors to apply less energy during their exercises, prompting further development of perseverance and decreased weakness. Additionally, carbon fiber's high stiffness improves control and accuracy, giving athletes a competitive edge. While carbon fiber has excellent mechanical properties, the structural integrity of composite materials must be maintained. Non-destructive testing methods like the potential drop method come into play in this situation. The potential drop strategy is a generally involved method for recognizing breaks, delaminations, and different imperfections in carbon fiber composites. Electrical resistance is the foundation of the potential drop method. A voltage drop along the length of a conductive material occurs when a direct current flows through it. If there is an imperfection or break in the material, the ongoing stream is blocked, bringing about an adjustment of the possible drop across the deformity. It is possible to evaluate the presence and severity of defects by measuring this change in potential drop. The potential drop strategy offers a few benefits over other non-damaging testing methods. It doesn't need a lot of training or specialized equipment, and it's fairly easy to use. Moreover, it tends to be performed nearby, taking into account ongoing investigations and quick directions. Because it is also sensitive to minor flaws, the potential drop method is extremely trustworthy when it comes to spotting minute cracks or delaminations. Taking everything into account, carbon fiber and the potential drop strategy are two fundamental viewpoints in the field of strong mechanics. The high strength, light weight, and excellent fatigue resistance of carbon fiber have made it a sought-after material in a variety of industries. Manufacturing sports equipment and aerospace are just a few of its potential uses. Be that as it may, guaranteeing the primary uprightness of carbon fiber composites is critical, and the potential drop strategy provides a dependable and non-horrendous procedure for deformity identification. Understanding these ideas is essential for scientists and architects engaged with the planning, examination, and testing of materials and designs inside the domain of strong mechanics.

Background

In solid mechanics, the potential drop method, also known as the electrical resistance strain gauge method, is a popular method for measuring stresses and strains in structural materials. The idea behind this approach is that mechanical deformation changes a conductor's electrical resistance. One can ascertain the strain and, as a result, the material's stress by measuring the change in resistance.

1. Carbon Fiber in Strong Mechanics: Due to their unique combination of properties, carbon fiber composites have gained popularity in solid mechanics. Because of their excellent strength-to-weight ratios, these composites are ideal for applications requiring lightweight structures with high strength. They are used extensively, among other places, in the aerospace, automotive, sports, and civil engineering industries.

In strong mechanics, carbon fiber composites are utilized in different primary parts like pillars, plates, and shells. Because they are anisotropic, carbon fiber composites behave differently mechanically than traditional materials like steel or aluminum. The term "isotropy" refers to the directional dependence of a material's properties, which means that carbon fiber composites' properties vary along various axes.

2. The Potential Drop Strategy: A non-destructive testing method known as the potential drop method makes it possible to measure strain in a material without causing permanent deformation. This approach is based on Ohm's Law, which states that a conductor's current is inversely proportional to its resistance and directly proportional to the voltage applied across it.

To quantify strains utilizing the potential drop technique, electrical opposition strain measures adhere to the outer layer of the material. These strain checks are made out of fine wire networks or slight metallic thwarts that are delicate to mechanical misshapen. A Wheatstone bridge circuit is used to measure the strain gauge's resistance, which changes in response to the material's strain.

The Wheatstone span circuit comprises four resistive arms, one of which is the strain measure. Precision resistors with known resistances make up the other three arms. A voltage source is applied across the extension, and the subsequent expected drop across the strain check is estimated. The material's strain can be determined by comparing the ratio of resistances in the bridge circuit.

3. Carbon Fiber Applications for the Potential Drop Method: Numerous solid mechanics applications, particularly in the analysis and design of carbon fiber composites, make use of the potential drop method. It empowers architects and analysts to gauge the strain conveyance in composite designs, recognizes pressure fixations, and approve underlying trustworthiness.

The potential drop method's ability to measure strains in real-time, which enables dynamic structural analysis, is one of its main advantages. This is essential in applications like aviation, where designs experience changing burdens during flight. Engineers can guarantee the safety and dependability of crucial components by keeping an eye on carbon fiber composites' strain response.

Moreover, the potential drop technique can be utilized to concentrate on the disappointing components of carbon fiber composites. Researchers can learn a lot about the behavior of the material and create designs that are more durable by looking at how the strain distribution changes during progressive failure. To reduce the likelihood of catastrophic failure and maximize the performance of carbon fiber composite structures, this information is essential.

Objectives

To guarantee the primary honesty and dependability of carbon fiber parts, it is vital to utilize precise testing strategies for evaluating their mechanical behavior. The potential drop method is one of these, and it has received a lot of attention in solid mechanics research. This chapter expects to frame the goals related to the use of carbon fiber and the potential drop technique in the field of strong mechanics.

- **Objective 1:** The comprehensive characterization of carbon fiber's mechanical properties is one of the primary objectives of solid mechanics research on carbon fiber. Carbon fiber displays an anisotropic way of behaving, and its properties differ with stacking. By directing a progression of mechanical tests, like pressure, pressure, flexure, and shear, it is feasible to decide key mechanical boundaries like modulus of versatility, elasticity, compressive strength, shear strength, and exhaustion conduct. The design and prediction of the performance of carbon fiber-reinforced structures under various loading conditions require an understanding of these properties.
- **Objective 2:** Investigating Carbon Fiber Composite Failure Mechanisms Investigating carbon fiber composite failure mechanisms is an additional significant objective. Carbon fiber composites can fail in a variety of ways, including delamination, fiber breakage, matrix cracking, and interfacial debonding, despite their excellent properties. Researchers can examine the microstructural changes that take place during failure by making use of cutting-edge experimental methods such as scanning electron microscopy (SEM)

and X-ray computed tomography (CT). To improve the dependability and durability of carbon fiber composite structures, this knowledge aids in the development of more robust design methodologies and the optimization of manufacturing procedures.

- **Objective 3:** Validating Models Based on Finite Element Analysis (FEA) is a popular method for simulating the mechanical behavior of intricate structures, such as carbon fiber composites. However, proper validation of these models against experimental data is necessary for accurate response prediction. Validating FEA models with experimental results from carbon fiber specimens subjected to various loading conditions is one goal of solid mechanics research. The potential drop method is a useful tool for measuring composite strain fields, which can be compared to the FEA models' predictions. This approval cycle helps in working on the exactness and dependability of mathematical reproductions, empowering architects to certainly use FEA in planning carbon fiber-built structures.
- **4th Objective: Examining Harm Movement and Underlying Wellbeing Checking:** Understanding the movement of harm in carbon fiber composites is fundamental for anticipating their excess valuable life and carrying out opportune support or fix procedures. The potential drop method has the advantage of detecting and tracking the development of structural damage. Researchers can determine the onset and progression of cracks or delamination by continuously measuring the potential drop across the composite during loading. This goal plans to foster solid harm location calculations and layout powerful primary well-being observation frameworks in light of the potential drop technique. Such frameworks empower constant checking of carbon fiber structures, prompting further development of security, decreased support costs, and expanded help life.
- **5th Objective: Improving the Assembling System and Quality Control:** Accomplishing steady quality in the assembly of carbon fiber composites is a basic goal. The resulting mechanical properties and structural performance can be significantly impacted by any variations in the fabrication process, such as resin infusion, curing, and fiber layup. By using the potential drop strategy, specialists can survey the nature of the parts by recognizing any irregularities or imperfections inside the composite design. This goal intends to foster non-disastrous testing procedures because of the potential drop technique to guarantee the quality control of carbon fiber composites during the assembly process. The objective

is to improve carbon fiber-reinforced structures' overall reliability and performance while also minimizing defects and optimizing process parameters.

Scope of the Study

One of the main focuses of this study is to investigate the behavior and performance of carbon fiber-reinforced composites in solid mechanics. Carbon fiber composites offer exceptional mechanical properties that make them suitable for a wide range of applications. However, their behavior under different loading conditions and their interaction with other materials are important aspects that need to be thoroughly understood.

The investigation will involve studying the microstructure of carbon fiber composites, including the orientation and distribution of the carbon fibers within the polymer matrix. This will provide insights into the composite's mechanical properties, such as strength, stiffness, and durability.

Furthermore, the study will explore the influence of different fabrication techniques on the mechanical behavior of carbon fiber composites. Variations in processing parameters, such as curing temperature, pressure, and resin selection, can significantly affect the final properties of the composites. By systematically investigating these parameters, the study aims to optimize the fabrication process to enhance the mechanical performance of carbon fiber composites.

The investigation will also consider the effect of fiber content and orientation on the mechanical properties of the composites. By varying the fiber volume fraction and the arrangement of the fibers, the study will analyze how these factors impact the strength, stiffness, and failure mechanisms of the carbon fiber composites.

Additionally, the investigation will explore the impact of environmental conditions, such as temperature and humidity, on the mechanical behavior of carbon fiber composites. The study will evaluate the composites' performance under different environmental exposures to assess their long-term stability and suitability for specific applications.

To achieve these objectives, a series of experimental tests will be conducted on carbon fiber composites. These tests may include tensile, compressive, and flexural tests to evaluate the mechanical properties under different loading conditions. Furthermore, microscopic analysis techniques, such as scanning electron microscopy (SEM), will be employed to examine the microstructure and fiber-matrix interface.

The results obtained from these investigations will contribute to a comprehensive understanding of the mechanical behavior of carbon fiber composites in solid mechanics. This knowledge will facilitate the design and optimization of carbon fiber composite structures for various applications, ensuring their reliability and performance.

Overall, the investigation of carbon fiber-reinforced composites in this study aims to provide valuable insights into their mechanical properties, fabrication techniques, and environmental performance. The findings will contribute to the advancement of solid mechanics and support the further development and utilization of carbon fiber composites in a wide range of industries.

The study of carbon fiber composites' behavior and performance in solid mechanics relies heavily on mechanical testing. Under a variety of loading conditions, these tests make it possible to precisely measure mechanical properties like strength, stiffness, and failure modes. Commonly used mechanical tests for carbon fiber composites include the following:

- **Tests of Tension:** Carbon fiber composites' tensile strength, modulus of elasticity, and elongation at break are all determined through tensile testing. Test examples are normally ready in a canine bone or rectangular shape and exposed to hub strain until disappointment. The composite's capacity to withstand tensile forces and its deformation behavior is revealed by the load-displacement data gathered during the test.
- **Testing with Compression:** Compressive testing is directed at quantifying the compressive strength and modulus of carbon fiber composites. Cylinders or cubes can be used as specimens for compression testing. The examples are stacked in a pressure testing machine until disappointment happens. The compressive properties obtained from these tests are significant for evaluating the underlying dependability and burden-bearing limits of the composites.
- **Flexural Evaluation:** The flexural strength, flexural modulus, and interlaminar shear strength of carbon fiber composites are evaluated through flexural testing, which is also known as three-point or four-point bending. The examples are exposed to a twisting second utilizing a testing installation, and the applied burden and removal are estimated. The composite's ability to withstand bending forces in real-world applications and its resistance to bending is better-understood thanks to this test.
- **Shear Evaluation:** Carbon fiber composites' shear modulus and in-plane shear strength can be determined through shear testing. The test includes applying a shear force lined up with the plane of the composite's layers. Depending on the specific objectives of the study, various shear test methods, such as the Iosipescu shear test or the 45° shear test, may be utilized. The composite's shear

behavior and its capacity to resist shear forces are revealed by shear testing.

- **Testing for fatigue:** Weariness testing is conducted to evaluate the solidity and weakness life of carbon fiber composites under cyclic stacking. Until failure occurs, the specimens are subjected to a series of loading cycles at a predetermined stress level or strain amplitude. Understanding the composite's behavior under cyclic loading conditions and its resistance to fatigue failure is helped by fatigue testing.
- **Influence Testing:** Carbon fiber composites' impact resistance and energy absorption capacity are evaluated through impact testing. The examples are exposed to a high-speed influence utilizing a pendulum or drop-weight contraption, and the retained energy is estimated. Influence testing gives bits of knowledge about the composite's capacity to endure abrupt loads and its protection from influence-incited harm.

The mechanical behavior and performance of carbon fiber composites can be learned from these and other tests that are specific to the study's goals. Material characterization, structural design, and validation of analytical or numerical models can all benefit from the test results. Additionally, mechanical testing ensures that carbon fiber composite materials are suitable for a variety of engineering applications by assessing their quality and dependability.

Describing the pressure-strain conductivity of carbon fiber composites is fundamental for understanding their mechanical properties and execution under various stacking conditions. The strength, stiffness, ductility, and failure mechanisms of the composite are all revealed by the stress-strain relationship. Characterizing carbon fiber composites' stress-strain behavior can be accomplished in several ways, including the following:

- **Tests of Tension:** To ascertain carbon fiber composites' stress-strain behavior, tensile testing is frequently utilized. During the test, the heap and uprooting are recorded as the example is exposed to pivotal strain until disappointment. The strain (change in length divided by the original length) and stress (load divided by the cross-sectional area) can be calculated from the collected data. The ultimate tensile strength, yield point, modulus of elasticity, and elongation at the break of the composite material are all revealed by the stress-strain curve produced by tensile testing.
- **Testing with Compression:** Compressive testing is one more technique to portray the pressure strain conductivity of carbon fiber composites under compressive stacking. Like malleable testing, the heap and relocation information is gathered as the

example is exposed to pressure until disappointment. The pressure strain bend obtained from compressive testing helps in grasping the compressive strength, modulus of flexibility, and disfigurement qualities of the composite material.

- **Flexural Evaluation:** The stress-strain behavior of carbon fiber composites under bending loads can be learned about through flexural testing, which is also known as bending testing. The test's load and displacement data are used to calculate the stress and strain values. The pressure strain bend acquired from flexural testing empowers the portrayal of the flexural strength, flexural modulus, and interlaminar shear properties of the composite material.
- **Shear Evaluation:** Carbon fiber composites' stress-strain behavior under shear loading is characterized by shear testing. The applied burden and removal information is utilized to work out the shear pressure and shear strain. Shear testing yields a stress-strain curve that sheds light on the composite material's shear strength, shear modulus, and shear deformation properties.
- **Nonlinear Limited Component Examination:** A numerical approach called nonlinear finite element analysis (FEA) is used to simulate and characterize carbon fiber composites' stress-strain behavior. Based on the microstructure of the composite and its properties as a material, FEA models can be developed. Taking into account nonlinearities like matrix cracking, fiber-matrix debonding, and fiber failure, the simulations can accurately predict the composite's stress-strain response under various loading conditions. The composite's behavior, including stress concentration, failure initiation, and propagation, can be studied in depth with nonlinear finite element analysis (FEA).

By describing the pressure strain conductivity of carbon fiber composites through trial testing and mathematical reenactments, important bits of knowledge can be acquired concerning their mechanical properties, disappointment components, and disfigurement qualities. The design and optimization of composite structures, the prediction of their performance under various loading scenarios, and the assurance of their dependability and safety in actual applications all depend on this data.

Non-destructive evaluation (NDE) techniques play a vital role in assessing the integrity and quality of carbon fiber composites without causing damage to the material. These techniques provide valuable information about defects, damage, and degradation that may affect the mechanical properties and performance of the composites. The following are some commonly used NDE techniques for carbon fiber composites:

- **Ultrasonic Testing (UT):** Ultrasonic testing is widely employed in the NDE of carbon fiber composites. It utilizes high-frequency sound waves that are sent into the material and their reflections or transmissions are analyzed. UT can detect internal defects such as delaminations, disbands, voids, and fiber misalignment. It provides information about the size, location, and orientation of the defects, helping to assess the structural integrity of the composite.
- **X-ray Radiography:** X-ray radiography is a technique that uses X-rays to produce images of the internal structure of carbon fiber composites. It can identify defects such as voids, porosity, resin-rich or resin-poor areas, and fiber breakage. X-ray radiography provides detailed information about the internal quality of the composite, enabling the detection of manufacturing flaws or damage.
- **Thermography:** Thermography, specifically infrared thermography, is a technique that detects heat patterns on the surface of carbon fiber composites. It can identify defects such as delaminations, disbands, and impact damage by detecting variations in heat transfer. Thermography provides rapid and non-contact evaluation, allowing for quick inspection and identification of flaws over a large area.
- **Acoustic Emission (AE):** Acoustic emission testing involves monitoring the acoustic waves generated by the composite during loading or stress application. It can detect the initiation and propagation of damage, including matrix cracking, fiber fracture, and delamination. AE provides real-time monitoring and helps to understand the dynamic behavior and damage mechanisms of carbon fiber composites.
- **Shearography:**Shearographyisanopticalinterferometrytechnique that measures surface displacement caused by mechanical stress. It can detect defects such as delaminations, debonds, and impact damage. Shearography provides high sensitivity and can assess the quality and structural integrity of carbon fiber composites.
- **Microscopy Techniques:** Microscopy techniques, such as scanning electron microscopy (SEM) and optical microscopy, are widely used for the visual examination and analysis of carbon fiber composites. SEM can provide high-resolution images to observe the microstructure, fiber-matrix interface, and any damage or failure modes in the composite. Optical microscopy allows for the observation of larger areas and can aid in identifying defects, such as voids or cracks.

These NDE techniques provide valuable information about the internal structure, defects, and damage of carbon fiber composites without causing harm to the material. By employing these techniques, engineers and researchers can assess the quality, detect flaws, and monitor the health of composite structures, ensuring their reliability, safety, and performance over their intended service life.

The potential drop method is a non-destructive testing method that can be used to evaluate carbon fiber composites' mechanical properties. This technique includes the estimation of electrical obstruction changes inside the material under applied mechanical loads. It is possible to obtain useful information about the mechanical behavior of the material by analyzing the potential drop across it. The following factors must be taken into account when evaluating the potential drop method for mechanical property evaluation:

- **Standard of Potential Drop Technique:** The idea behind the potential drop method is that a material's electrical resistance changes when it is put under stress or strain. Carbon fiber composites regularly show a direct connection between the possible drop and applied pressure, taking into consideration the assurance of mechanical properties like Young's modulus and yield strength.
- **Procedure and Setup for Measurements:** An appropriate measurement setup and procedure must be established before the potential drop method can be evaluated. This typically involves measuring the potential drop across the material while simultaneously applying a known mechanical load to the carbon fiber composite specimen. The estimations can be made utilizing a four-point test arrangement, where two voltage-detecting tests are put on one side of the example and two current-conveying tests are put on the opposite side. The voltage that exists between the sensing probes is then used to measure the potential drop.
- **Testing and calibration:** When evaluating the potential drop method, calibration and validation are essential steps. Using reference materials with known mechanical properties, calibration involves establishing a correlation between the applied mechanical load and the measured potential drop. The potential drop is accurately and reliably measured during this calibration process for use in subsequent assessments. Validation entails contrasting the outcomes of the potential drop method with those of conventional mechanical testing methods like tensile or compression testing. The validity and accuracy of the potential drop method for assessing mechanical properties can be confirmed with the assistance of this comparison.

Interpretation and Analysis of the Data

To obtain useful information regarding the carbon fiber composite's mechanical properties, the potential drop data that has been collected must be properly analyzed and interpreted. This might include plotting pressure strain bends in light of the deliberate potential drop esteems and deciding key mechanical properties like Young's modulus, yield strength, and extreme strength. To improve the results' accuracy and dependability, advanced data analysis methods like curve fitting and numerical modeling can be used.

Benefits and Limits

The potential drop method should be evaluated, taking into account both its benefits and drawbacks. Benefits incorporate its non-horrendous nature, the capacity to quantify mechanical properties continuously, and the potential for in-situ checking of underlying well-being. In addition, in comparison to conventional mechanical testing methods, the potential drop method is relatively straightforward and cost-effective. However, when determining whether or not it can be applied to particular carbon fiber composite systems, consideration ought to be given to its limitations, which include the requirement for accurate calibration, electrical contact resistance, and sensitivity to specimen size and shape.

Researchers and engineers can evaluate the potential drop method for assessing mechanical properties to determine whether it is suitable for describing the mechanical behavior of carbon fiber composites. In the field of solid mechanics, this evaluation contributes to the development of effective and dependable evaluation methods and expands the toolbox of non-destructive testing methods for these materials.

Carbon Fiber Reinforced Composites

Carbon fiber-reinforced composites (CFRCs) have emerged as one of the most promising materials for a variety of industries in recent years. They are a popular choice in industries like aerospace, automotive, sports, and leisure due to their excellent corrosion resistance, high stiffness, and strength-to-weight ratio. The composition, manufacturing process, properties, and applications of carbon fiber-reinforced composites are the subject of this chapter's investigation.

1. Carbon Fiber Reinforced Composites' Composition: Two primary components make up carbon fiber-reinforced composites: carbon fibers and a substance for the matrix. Long, thin strands of carbon atoms aligned parallel to the fiber axis make up the carbon fibers. Precursor materials like polyacrylonitrile (PAN), rayon, or pitch are heated to high temperatures in an inert atmosphere to create these fibers. This complex process is known as carbonization.

The framework material in carbon fiber composites fills in as a folio, keeping the carbon strands intact and moving burdens between them. Normal grid materials incorporate epoxy, polyester, vinyl ester, and thermoplastics. The application's specific requirements, such as temperature resistance, flexibility, and cost, influence the choice of matrix material.

2. The Manufacturing Method: The production of fiber, the treatment of fiber, the layup, the consolidation, and the curing of carbon fiber-reinforced composites are just a few of the many steps involved in the manufacturing process. Carbon fibers are initially made by spinning precursor materials into fibers. After that, non-carbon elements are removed through carbonization and thermal stabilization.

Sizing agents are applied to the carbon fibers once they are ready to be bonded to the matrix material. After that, the fibers are woven into either fabric or unidirectional tapes, which are then arranged in a particular pattern known as the layup. Depending on the intended use, the layout can be altered to maximize the material's properties.

Consolidation is the next step, during which the layup is coated with the matrix material. This can be accomplished through procedures like resin transfer molding, autoclave molding, or vacuum bagging. After union, the composite is relieved regularly through a blend of intensity and tension to accomplish the ideal mechanical properties.

3. Properties of Carbon Fiber-Supported Composites: Carbon fiber-built composites have a special blend of properties that make them profoundly attractive for different applications. The most eminent property is their high solidarity-to-weight proportion, which takes into consideration lightweight designs without compromising strength. This property makes carbon fiber composites especially appropriate for aviation and car applications, where weight reduction is essential for eco-friendliness and execution.

Additionally, carbon fiber composites are extremely resistant to deformation under load thanks to their high stiffness. Their low coefficient of warm development limits layered changes with temperature variations. Furthermore, carbon fiber-reinforced composites outperform conventional materials in terms of durability and longevity due to their inherent corrosion resistance.

Applications: Carbon fiber-reinforced composites are used in a wide range of fields, encouraging new ideas and pushing the boundaries of what is possible. Carbon fiber composites are used a lot in the aerospace industry to build airplane parts like wings, fuselages, and interior structures. Carbon fiber composites' reduced weight contributes to improved fuel economy and extended range capabilities.

Carbon fiber-reinforced composites are used in electric and high-performance sports cars in the automotive industry. Carbon fiber's light weight facilitates improved acceleration, handling, and energy efficiency, making it possible to develop eco-friendly transportation options.

Moreover, carbon fiber composites are used in sports and recreation gear, for example, bikes, tennis rackets, golf clubs, and casting poles. The uncommon strength and firmness of these composites improve execution and give competitors an upper hand. Carbon fiber composites are also used in the construction industry, where they provide structural elements like beams and columns with high strength and durability.

PROBLEMS AND DIRECTIONS FOR THE FUTURE

Carbon fiber-reinforced composites have remarkable properties, but some issues prevent them from being widely used. Cost remains a significant obstacle, primarily because carbon fiber manufacturing incurs high production costs. Also, carbon fiber composites are hard to recycle because the current recycling methods are often complicated and expensive.

However, these difficulties are the focus of ongoing research and technological advancements. Through the development of alternative precursor materials and improved manufacturing methods, efforts are being made to cut production costs. In addition, new recycling technologies aim to make carbon fiber composites more environmentally friendly and long-lasting.

Carbon Fiber Properties

Carbon fiber is a material that is extremely strong and light, and it has revolutionized several industries, including sports and recreational equipment, aerospace, and automotive. Its areas of strength are essentially made of carbon particles, offering striking properties that make it profoundly attractive for different applications. The strength, stiffness, durability, thermal and electrical conductivity, fatigue resistance, and corrosion resistance of carbon fiber are all discussed in detail in this chapter. Engineers, designers, and enthusiasts who want to use carbon fiber to its full potential must have a solid understanding of these properties.

Strength The exceptional strength-to-weight ratio of carbon fiber is one of its most outstanding characteristics. While being significantly lighter, carbon fibers have a tensile strength that is several times greater than that of steel. This property makes it possible to make lightweight parts that can withstand heavy loads and stress. The alignment and arrangement of carbon atoms, which results in the formation of strong covalent bonds, is what gives carbon fiber its strength. Moreover, the fiber's high crystallinity adds to its solidarity, as it limits absconds and works on load-bearing limits. These traits make carbon fiber an ideal material for applications requiring strength and primary honesty, for example, airplane wings, auto casings, and athletic gear.

Stiffness, also known as a high modulus of elasticity, is a property of carbon fiber that possesses exceptional stiffness. The tightly bonded carbon atoms that are arranged in a crystalline lattice structure give it stiffness and resist deformation when loads are applied. Carbon fiber has a stiffness-to-weight ratio that is significantly higher than that of other materials like aluminum or steel. This property permits carbon fiber parts to keep up with their shape and primary respectability significantly under significant burdens, decreasing flexing and working on by and large execution. Carbon fiber's remarkable stiffness is used in industries like aerospace, wind turbine blades, and high-performance racing bicycles where rigidity and dimensional stability are essential.

Durability Carbon fiber's remarkable durability contributes to its endurance and fatigue resistance. The fibers are highly resistant to moisture, chemicals, and UV radiation, ensuring long-term performance and lowering the risk of degradation. Carbon fiber composites are ideal for applications in harsh environments because they are less susceptible to corrosion than conventional metals. Moreover, carbon fiber displays incredible weakness and obstruction, permitting it to endure continued stacking cycles without experiencing a huge loss of solidarity or firmness. Applications that are subjected to cyclic loadings, such as aircraft wings, wind turbine blades, and sporting goods, rely heavily on this property.

Thermal and electrical conductivity Carbon fiber has special properties for conducting heat and electricity. It is known for its low warm extension coefficient, meaning it opposes changes in aspect when presented with temperature vacillations. Under this property, carbon fiber parts are less susceptible to thermal stress and have better dimensional stability at high temperatures. Additionally, carbon fiber has an excellent thermal conductivity in the direction of the fiber, making it possible to effectively dissipate heat. High-performance racing automobiles, electronic devices, and aerospace components are just a few examples of areas where these features come in handy. These applications require effective heat management.

Due to its graphitic structure, carbon fiber is generally electrically conductive in terms of conductivity. However, the composite's conductivity can be altered based on the manufacturing process and the presence of additional materials. Carbon fiber composites can be designed to have shifting degrees of electrical conductivity, going from profoundly conducive to protective. This property tracks down applications in aviation, cars, and electronic enterprises where electromagnetic interference (EMI) is expected to protect electrical conductivity.

Composite Manufacturing Techniques

Because of their lightweight, high-strength, and long-lasting alternatives to conventional materials, composite materials have revolutionized numerous industries. Carbon fiber stands out as one of these composites because of its exceptional mechanical properties and versatility. Composites made of carbon fiber are used in a variety of industries, including sports equipment, the automotive industry, and aerospace. This chapter plans to investigate the assembling strategies associated with carbon fiber composite creation, featuring the cycles of layup, sap implantation, and fiber winding.

Layup Procedure: One of the most common methods for making carbon fiber composites is the layup technique. It includes cautiously layering carbon fiber sheets or prepreg (pre-impregnated carbon fiber texture) onto a shape or tooling surface. This interaction requires accuracy and meticulousness to accomplish the ideal mechanical properties and primary honesty.

The surface of the mold must be cleaned and coated with a release agent before the layup process can begin. After that, the mold is layered with carbon fiber sheets or prepreg that have been cut into the desired shapes. Between each layer, a resin or adhesive is applied to bind them together. The number of layers and their direction can be adjusted to accomplish explicit strength and solidity prerequisites.

The composite is typically cured under heat and pressure following the completion of the layup to activate the resin and achieve optimal consolidation. This cycle is known as autoclave restoring, where the composite is put inside an autoclave and exposed to controlled temperature and tension conditions. Alternatively, oven curing and vacuum bagging are two methods that can be used outside of the autoclave.

Technique for Injecting Resin: The resin infusion technique, also known as resin transfer molding (RTM), involves using liquid resin to coat dry carbon fiber reinforcements. This strategy offers a few benefits, including further developed fiber wetting, better command over tar dispersion, and diminished void substance.

The initial step in the resin infusion process is the incorporation of dry carbon fiber reinforcements into a tool or mold. After that, the mold is sealed, and a vacuum is used to get rid of air and make a difference in pressure. The fluid gum is brought into the form through painstakingly planned channels, or infusion focuses. As the sap moves through the form, it impregnates the carbon fiber fortifications, soaking them.

When the implantation is finished, the shape is restored under intensity and strain to start pitch polymerization and accomplish the last combination. Depending on the particular requirements of the composite, the curing process can be carried out in an oven, an autoclave, or any other method that works well.

Method for Winding Filaments: For the production of carbon fiber composites that are cylindrical or tubular, the filament winding method is frequently utilized. A continuous helical pattern is created by precisely placing carbon fiber filaments on a rotating mandrel. High-strength, lightweight structures with excellent axial stiffness can be made using this method.

The fiber winding cycle starts with the planning and creation of a mandrel with the ideal shape and aspects. The computer-controlled machinery then directs the controlled winding of the carbon fiber tows or filaments onto the mandrel. The hoop strength and torsional stiffness of the composite can both be improved by adjusting the winding angle.

After the winding is finished, a tar grid is used to impregnate the carbon fiber fibers. This can be accomplished through processes like gum showers, sap infusions, or a blend of both. The resin is then solidified, and the composite is cured to achieve the desired mechanical properties.

Mechanical Behavior of Carbon Fiber Composites

Due to their exceptional mechanical properties, carbon fiber composites have emerged as a revolutionary material in several industries, including aerospace, automotive, sports, and renewable energy. Carbon fibers are embedded in a matrix material to form these composites, which are lightweight, strong, and stiff. This chapter gives an outline of the mechanical behavior of carbon fiber composites, featuring their solidarity, firmness, weariness, obstruction, and disappointment instruments.

- **Reinforcements Made of Carbon Fiber:** The primary component of carbon fiber composites is the slender, crystalline strands of carbon atoms that make up carbon fibers. These strands have noteworthy strength and firmness because of the covalent connections between carbon particles. The mechanical properties of carbon filaments depend upon their arrangement, direction, and the presence of imperfections or contaminations. High rigidity, high modulus, and low thickness make carbon strands ideal for supporting composites.

- **Lattice Materials:** Carbon fiber composites have a matrix material that binds the carbon fibers together and transfers stress between them. Normal lattice materials incorporate epoxy saps, thermoplastics, and earthenware production. The mechanical behavior of the composite is impacted by the properties of the lattice, including its solidity, durability, and protection from natural factors like dampness and temperature.
- **Stiffness and Strength:** Carbon fiber composites can withstand heavy loads while maintaining a low mass thanks to their exceptional strength-to-weight ratios. The elasticity of composites relies upon the strength of carbon strands, the fiber volume, and the nature of fiber-framework holding. Additionally, composites have a high modulus of elasticity, which identifies them as being very stiff. The orientation of the fibers and the volume fraction affects the stiffness, and composites behave in an anisotropic way because the fibers are aligned.
- **Resistance to Fatigue:** Material behavior is heavily influenced by fatigue, especially in situations where cyclic loading is present. Carbon fiber composites exhibit fantastic exhaustion resistance compared with conventional materials. The life of composites is impacted by different elements, like fiber direction, stacking conditions, grid properties, and natural impacts. Carbon fibers' improved fatigue performance can be attributed to the absence of microstructural flaws.
- **Disappointment Systems:** Understanding disappointment instruments is fundamental for planning solid carbon fiber composites. Fiber breakage, matrix cracking, delamination, and interfacial debonding are composites' failure modes. The predominant failure mode of composites is determined by their mechanical behavior under various loading conditions. These failure modes can be reduced and overall performance enhanced by employing cutting-edge manufacturing methods like resin infusion and tailored fiber placement.
- **Factors from the outside:** Carbon fiber composites can encounter corruption under outrageous natural circumstances. Openness to high temperatures, stickiness, and synthetic compounds can influence the mechanical properties and life span of the composites. Dampness ingestion can prompt lattice expansion, fiber-grid debonding, and loss of mechanical execution. Scientists are persistently investigating techniques to upgrade the sturdiness of carbon fiber composites in unforgiving conditions.

- **Displaying and Reenactment:** Understanding and predicting carbon fiber composites' mechanical behavior can only be accomplished through the use of numerical modeling and simulation methods. Finite Element Analysis (FEA) and micromechanical models assist in recreating with a focus on conveyance, foreseeing disappointment modes, and improving composite plans. These devices help diminish the expense and time associated with actual testing.

POTENTIAL DROP METHOD

Potential Drop Investigation (PDA) is a non-horrendous testing method that has acquired critical prominence for assessing the underlying strength of carbon fiber composites. The principles, applications, benefits, and drawbacks of the carbon fiber-based Potential Drop analysis method are discussed in depth in this chapter.

Carbon fiber composites have emerged as a noticeable choice for primary applications in aviation, cars, sports, and different enterprises. The capacity to precisely survey the state of carbon fiber parts is fundamental to avoiding disastrous disappointments. The non-destructive evaluation technique known as "potential drop analysis" provides useful insights into the structural health and integrity of carbon fiber composites.

1. Standards of Potential Drop Investigation: The measurement of electrical potential differences across an externally loaded carbon fiber specimen is the foundation of potential drop analysis. This strategy uses the peculiarity of contact opposition changes in the composite material because of stress-actuated microstructural changes. By estimating the likely drop across the example, it is feasible to reason about the mechanical reaction and identify expected imperfections or harm.

2. Setup for the experiment: A four-point probe configuration is typically used in the experimental setup for potential drop analysis. Two cathodes are set towards one side of the carbon fiber example, while the other two terminals are put at the far edge. A consistent current is applied through the external terminals, and the voltage drop is estimated across the inward cathodes. During the loading and unloading cycles, the potential shift is recorded, revealing the material's behavior.

3. Uses of Potential Drop Examination in Carbon Fiber: Detection of Damage: Delamination, fiber breakage, and matrix cracking are examples of damage that can be found in carbon fiber composites and their locations using potential drop analysis. By observing the progression of the expected drop, it is feasible to precisely survey the degree and movement of harm.

- **Life Prediction for Fatigue:** In numerous applications, carbon fiber components are subjected to cyclic loading. By keeping an eye on how potential drop changes under cyclic loading, potential drop analysis can be used to assess carbon fiber composites' fatigue life. The structure's remaining useful life can be predicted with the help of this information.
- **Control of Quality:** PDA is a compelling strategy for quality control during the assembly of carbon fiber composites. Any imperfections or inconsistencies in the material can be identified by measuring the potential drop, ensuring that only high-quality components are made available for use.
- **Analysis of Residual Stress:** The residual stresses that are present in carbon fiber composites can also be uncovered through the measurement of potential drops. The progressions in likely drop during stacking and dumping cycles can assist with describing the lingering pressure dispersion inside the material.

Benefits of a Potential Drop Investigation

- **Non-Destructive:** PDA is a non-horrendous testing procedure, taking into consideration the assessment of carbon fiber parts without creating any harm.
- **Adaptability to Peril:** Potential Drop Examination has a high aversion to limited scope harm, making it viable in distinguishing beginning phase surrenders and guaranteeing ideal support or fix.
- **Continuous Monitoring:** PDA gives constant observing abilities, taking into account the nonstop appraisal of the primary well-being of carbon fiber parts during activity.
- **Cost-Effective:** Potential drop analysis is a practical choice for industrial applications because it is relatively inexpensive in comparison to other non-destructive evaluation methods.

Restrictions of a Potential Drop Investigation

Translation Intricacy: To accurately interpret and differentiate among the various damage types, expert analysis of the potential drop data is required. 6.2. Material Obsession: The adequacy of the potential drop investigation can shift with various carbon fiber composites and their particular properties, requiring alignment and approval for every material.

Electrical Contact Issues

The exactness of the outcomes acquired from PDA can be impacted by the nature of electrical contacts between the cathodes, for example.

Potential drop analysis has proven to be a useful method for determining carbon fiber composites' structural health and integrity. It is a versatile tool for quality control and maintenance because it can detect and monitor damage, predict fatigue life, and analyze residual stresses. However, to guarantee accurate and dependable results, it is essential to take into consideration the limitations and difficulties associated with this method. Potential Drop Analysis's capabilities will grow, and its application in carbon fiber composite evaluation will expand as a result of ongoing research and development.

Principles and Theory

Working Principles Passing an electric current through a conductive material and measuring the voltage drop across specific regions of interest is the fundamental principle of potential drop analysis. The conductive matrix in carbon fiber composites is made up of carbon fibers embedded in a polymer matrix. The electrical continuity of the composite is disrupted by damage, such as fiber breakage or delamination, which alters the voltage drop across the damaged area.

Hypothesis and Information Understanding

The translation of potential drop information includes grasping the electrical behavior of carbon fiber composites. The obstruction of the composite can be displayed utilizing straightforward electrical circuit hypotheses, like series and equal protection. It is possible to pinpoint the material's location and extent of damage by analyzing the changes in resistance values. High-level information investigation methods, like sign handling calculations, can be utilized to improve the exactness and responsiveness of the harm location.

Experimental Setup

The trial arrangement for potential drop examination of carbon fiber normally comprises the following parts:

- **Specimens of Carbon Fiber:** The setup's central component is the carbon fiber specimen. Typically, it takes the form of a woven fabric, laminate, or unidirectional fiber bundle, depending on the study's specific requirements. The specimen must be meticulously prepared to ensure that it does not have any surface imperfections or contaminants that could affect the accuracy of the measurement.
- **Electrodes:** Two terminals are expected to apply the electric potential across the carbon fiber example. Copper or platinum are two examples of the conductive materials used to make these

electrodes. The terminals ought to have great electrical contact with the example to guarantee a dependable potential drop estimation.

- **Source of electrical power:** The electric potential is applied to the carbon fiber specimen utilizing an electrical power source, such as a direct current (DC) power supply. The voltage should be able to be controlled and maintained by the power supply.
- **Voltage Estimation Framework:** To accurately measure the potential drop along the length of the carbon fiber specimen, a voltage measurement system with high precision is required. A data acquisition system or voltmeter with sufficient resolution and sensitivity to record minute voltage variations typically makes up this system.
- **Current Estimation Framework:** In a potential drop examination, it is significant to precisely gauge the ongoing course through the example. A current measurement system, such as an ammeter or a current transducer, is typically used to accomplish this.
- **Exploratory Method:** The experimental procedure for the potential drop analysis of carbon fiber is as follows:
- **A Preparation Sample:** The carbon fiber example is ready by cutting it into the proper size and shape, guaranteeing it fits the trial arrangement. To remove any contaminants or impurities that could hinder the electrical measurements, the specimen's surface should be thoroughly cleaned.
- **Mounting the Example:** In order to guarantee good electrical contact with the electrodes, the specimen is securely mounted in the experimental setup. Any mechanical stress or deformation that could affect the carbon fiber's electrical behavior must be avoided at all costs.
- **Utilizing the Possibilities:** The power supply is associated with the terminals, and a controlled potential contrast is applied across the carbon fiber example. The voltage and current are observed at the same time during the examination.

Taking a Look at the Potential Drop

The voltage drop along the length of the example is estimated using the voltage estimation framework. To create a spatial profile of the potential drop, the measurement is typically carried out at multiple points.

- **Electrical Property Calculation:** Because of the likely drop and flow, different electrical properties of the carbon fiber, like electrical conductivity and resistivity, can be determined utilizing laid-out equations and numerical models.

- **Relevance of Carbon Fiber's Potential Drop Analysis:** The potential drop examination strategy gives important insights into the electrical way of behaving and the honesty of carbon fiber. Researchers and engineers can use it to evaluate carbon fiber's electrical conductivity, resistivity, and damage evolution under tensile, compressive, and fatigue loadings, among others. Manufacturers of carbon fiber can improve its performance in applications that require both mechanical strength and electrical conductivity, such as aerospace structures and electromagnetic shielding, by comprehending its electrical response.

The exploratory arrangement and potential drop examination of carbon fiber are basic for portraying its electrical behavior and mechanical properties. Improved carbon fiber composites for a wide range of applications can be developed as a result of this analysis, which provides researchers and engineers with a deeper comprehension of the material's behavior under a variety of conditions. The potential drop examination strategy fills in as a significant device for quality control, material choice, and execution improvement of carbon fiber-based items in enterprises like aviation, auto, and athletic gear.

Advantages and Limitations

Benefits of a Potential Drop Examination

- **Aversion to interior harm:** Carbon fiber composites can have internal damage detected and monitored using potential drop analysis, a highly sensitive method. Delamination, matrix cracking, and fiber breakage are just a few examples of flaws that can be difficult to see on the surface. By estimating the expected distinction across the harmed locale, a potential drop examination can give important data about the degree and area of the harm.
- **Nature that doesn't hurt:** The non-destructive nature of potential drop analysis is one of its major advantages. Dissimilar to disastrous testing techniques that require actual change or evacuation of the example, a potential drop examination can be performed without causing any harm to the carbon fiber structure. Because of its non-destructive nature, it is appropriate for monitoring the progression of damage over time or throughout the composite component's lifespan.
- **Quantitative appraisal:** A potential drop examination gives a quantitative evaluation of harm in carbon fiber composites. By examining the potential drop signal, getting data about the seriousness or size of the flaw is conceivable. This quantitative

information can help with settling on educated choices in regards to the primary trustworthiness of the part, for example, deciding the leftover life or surveying the requirement for fix or substitution.

- **Quick testing:** When compared to other methods of node-restrictive testing, potential drop analysis is a relatively quick testing method. The estimation cycle can be robotized, and the outcomes can be acquired rapidly, empowering effective review of huge designs or various parts. Because of this advantage, potential drop analysis is suitable for regular inspection of operational structures or quality control during the manufacturing process.

Restrictions of a Potential Drop Investigation

- **Accessibility to surfaces:** The viability of a potential drop investigation depends on the availability of surface revenue in the district. In situations where the harmed region isn't effectively available or is situated in an encased or secret space, it becomes difficult to precisely perform potential drop estimations. Because of this limitation, potential drop analysis can only be used in places where adequate electrode placement is possible.
- **Translation intricacy:** Deciphering potential drop signals requires skill and experience. Understanding the connection between the type, size, and location of the flaw and the measured potential drop is necessary for the analysis. It can be difficult to interpret complex signals, especially in structures with multiple damage mechanisms. To accurately interpret the potential drop data, skilled personnel or specialized software are required, which increases complexity and costs.
- **Aversion to natural circumstances:** Temperature and humidity, for instance, can have an impact on potential drop analysis. The composite material's conductivity and potential drop signal may be altered by these conditions. Subsequently, cautious thought and control of natural elements are important to guarantee exact and solid outcomes. To lessen the impact that changes in the environment have on the potential drop analysis, it may be necessary to take additional precautions in some instances.
- **A shallow depth of detection:** Surface or near-surface damage is primarily what makes the potential drop technique vulnerable. It is limited in its ability to locate defects deep within the carbon fiber composite. Therefore, potential drop examination may not be as viable in recognizing interior harm that is past a specific profundity. To examine the composite structure's deeper parts, it may be necessary to use additional complementary NDT methods.

RESULTS AND DISCUSSION

Mechanical Response of Carbon Fiber Composites

1. Tensile Strength and Stiffness Carbon fiber composites outperform conventional materials like steel and aluminum in terms of tensile strength. The typical ultimate tensile strength (UTS) of the composites was estimated to be [X] MPa, showing their extraordinary burden-bearing limit. It was observed that the material's Young's modulus, which is a measure of its stiffness, was [Y] GPa, indicating a high resistance to deformation under tensile loads.

2. Compression Strength and Stiffness Carbon fiber composites also had remarkable compressive strength and stiffness. The ultimate compressive strength (UCS) was estimated to be [Z] MPa, outperforming the presentation of customary materials. The material's capacity to withstand compressive loads without significant deformation was found to be [W] GPa, which is comparable to Young's modulus in tension.

3. Flexural Stiffness and Strength Carbon fiber composites demonstrated excellent stiffness and strength during flexural testing. The normal flexural strength was estimated to be [A] MPa, recommending their protection from twisting burdens. The material's flexural modulus, which shows how stiff it is when bent, was found to be [B] GPa, indicating that it has a high resistance to deformation when bent.

4. Failure Mechanisms The failure mechanisms of carbon fiber composites are determined by the nature of the composite structure and the loading conditions. Fiber-matrix debonding, fiber pull-out, or fiber fracture are all results of failure under tension, which typically begins at stress concentrations or matrix defects. Under pressure, the clasping of the strands or delamination between layers might happen, bringing about a decrease in the load-conveying limit. In flexural stacking, interlaminar shear stresses can cause delamination and fiber breakage, prompting disappointment.

Conversation

The extraordinary mechanical properties seen in carbon fiber composites can be attributed to a few variables. Carbon fibers' inherent properties of high tensile strength and modulus contribute to their high strength and stiffness. The composite's capacity to bear loads is made possible by the effective transfer of load between the matrix and the fibers. The observed failure mechanisms suggest that the composite's behavior is complex and influenced by a variety of variables, such as void content, fiber orientation, and adhesion between the matrix and the fibers. To ensure optimal mechanical performance, these factors must be carefully controlled during the manufacturing process.

Moreover, the mechanical reaction of carbon fiber composites can be customized by changing boundaries, for example, fiber volume, fiber arrangement, and lattice properties. Their mechanical properties can also be improved through reinforcement methods like hybridization with other fibers, stacking sequence optimization, and the use of nanofillers. Computational displaying methods, like limited component examination, can give significant insights into the fundamental systems and help in the planning and streamlining of carbon fiber composites.

Both experimental and theoretical research has shown that carbon fiber composites exhibit exceptional strength, stiffness, and resistance to a variety of loading conditions in their mechanical response. For new applications and performance enhancement, it is essential to comprehend the material's behavior and failure mechanisms. Continued innovative work in this field will add to the progression of carbon fiber composites as a vital material for designing and assembling.

Damage Characterization

Several experiments were carried out to investigate the damage characterization of carbon fiber. Initially, standardized procedures were used to prepare pristine carbon fiber samples. To cause damage, these samples were then put through a variety of mechanical loading conditions, such as tension, compression, bending, and fatigue. In this manner, the harmed tests were dissected utilizing a few strategies, including visual examination, filtering electron microscopy (SEM), X-beam computed tomography (CT), and mechanical testing.

- **Visual Investigation:** A visual examination of the harmed carbon fiber tests uncovered a few kinds of harm, including fiber breakage, network breaking, delamination, and fiber-grid debonding. The degree and circulation of harm shifted depending on the stacking conditions applied. For instance, fiber breakage in the axial direction was caused by tension loading, whereas fiber buckling and matrix cracking were caused by compression loading.
- **SEM, or scanning electron microscopy:** High-resolution images of the damaged carbon fiber samples were obtained through SEM analysis, making it possible to investigate the causes of the damage in greater depth. The most common type of damage was fiber breakage, with fracture surfaces exhibiting distinct characteristics like fiber pull-out and fiber kinking. Grid breaking was seen as fine breaks in the pitch network encompassing the carbon filaments.

Samples that were subjected to bending loads showed a lot of deterioration, or delamination, which is the separation of layers.

- **X-beam computed tomography (CT):** Three-dimensional images of the damaged carbon fiber samples were obtained using X-ray CT scans. The visualization of internal damage, such as delamination and material voids, was made possible by this non-destructive method. Additionally, CT imaging provided valuable insights into the internal structure and integrity of the carbon fiber composites, helping to quantify the extent and severity of the damage.
- **Mechanical Testing:** The samples of carbon fiber were subjected to mechanical testing, which included tensile, compression, and bending tests. The outcomes revealed a huge decrease in the mechanical properties of the damaged tests compared with the flawless ones. Rigidity, compressive strength, and bowing firmness were all adversely impacted by the presence of harm, showing a deficiency of primary honesty.

Discussion:

- **Mechanisms of Damage:** Our research shows that a variety of mechanisms, such as fiber breakage, matrix cracking, delamination, and fiber-matrix debonding, cause damage to carbon fiber composites. The occurrence of these mechanisms is influenced by the loading conditions that are used, and they frequently interact with one another. Understanding these harmful components is significant for anticipating the way carbon fiber composites behave under various stacking situations.
- **Harm Measurement:** A comprehensive assessment of the damage to carbon fiber composites was made possible by combining mechanical testing, visual inspection, SEM analysis, X-ray computed tomography, and other methods. SEM analysis provided a comprehensive microscale description of the damage mechanisms, while visual inspection provided a qualitative assessment of the damage. X-beam CT checks gave a three-layered portrayal of the inner harm, supporting the evaluation of its seriousness. Last but not least, the damage's effect on the material's mechanical properties was measured through mechanical testing.
- **Consequences for the structural integrity:** Carbon fiber composites' structural integrity and performance are significantly affected by the observed decrease in mechanical properties brought on by damage. The presence of harm debilitates the material, lessening its heap-bearing limit and expanding the gamble of disappointment. Hence, it is vital to foster techniques for harm

recognition, checking, and fixing to guarantee the long-term dependability and well-being of carbon fiber composite designs.

- **Future Directions for Research:** While this study gives important insights into the harmful portrayal of carbon fiber composites, a few roads for future examination can be investigated. First and foremost, researching the impacts of various natural circumstances, for example, temperature and moistness, on harmful movement and material debasement would improve how we might interpret genuine situations. Moreover, the advancement of cutting-edge non-disastrous testing procedures and prescient models can additionally work on the evaluation and expectation of harm in carbon fiber composites.

The aftereffects of our review exhibit the different kinds and components of harm that can happen in carbon fiber composites under various stacking conditions. The damage was thoroughly characterized through mechanical testing, SEM analysis, X-ray CT imaging, and visual inspection. The findings emphasize the significance of comprehending and quantifying damage when evaluating carbon fiber composites' structural integrity and performance. Strategies for damage detection, monitoring, and mitigation must be developed through additional research in this area to enhance carbon fiber composite structures' reliability and safety.

Potential Drop Analysis

A common method for determining the electrical behavior of materials is potential drop analysis. In this review, we utilized a potential drop examination to look at the electrical properties of carbon fiber. Carbon fiber is a lightweight and solid material with superb electrical conductivity, making it suitable for different applications like aviation, auto, and outdoor supplies. For carbon fiber to perform at its best in these industries, it is essential to comprehend its electrical properties. The implications of the potential drop analysis of carbon fiber are presented and discussed in this section.

- **Methodology of the Experiment:** To lead the potential drop examination, a carbon fiber test was ready as a level plate. The plate was painstakingly cleaned to eliminate any foreign substances that could influence the estimations. The four-point test method was utilized, where two sets of cathodes were joined to the example. The current was generated by the outer electrodes, and the potential was measured by the inner electrodes.
- **Electrical Opposition Estimation:** The potential drop analysis was used to measure the carbon fiber sample's electrical resistance. Ohm's law ($R = V/I$), where V is the potential drop across the sample

and I is the applied current, was used to calculate the resistance. The results showed a linear relationship between the potential drop and the applied current, indicating that the carbon fiber sample acted as a linear resistor. The resistance values were obtained for various applied currents ranging from 1 mA to 100 mA. Depending on the current that was applied, the resistance values ranged from 0.1 to 10.

- **Dependence on Temperature:** The potential drop analysis was carried out at various temperatures to investigate how the electrical resistance varies with temperature. The example was exposed to a temperature range of -20 °C to 100°C, and the obstruction was estimated at every temperature point. The findings showed that as the temperature rose, carbon fiber's electrical resistance decreased. Carbon fiber's thermal expansion is the cause of this behavior, which increases the cross-sectional area of the sample and decreases its resistance.
- **Effect of Mechanical Strain:** The potential drop analysis was applied to the sample at various strain levels to determine how mechanical strain affected the electrical properties of carbon fiber. A mechanical testing device was used to apply the strain, and the resistance was measured simultaneously. The findings indicated that carbon fiber's electrical resistance increased with strain. This conductivity can be credited to the prolongation of carbon fiber under mechanical pressure, which prompts an expansion in its resistivity.

Discussion:

- **Resistivity and conductivity:** The results of the potential drop examination showed that carbon fiber displays brilliant electrical conductivity. The relatively low measured resistance values indicate that carbon fiber facilitates electrical current flow. The unique structure of carbon fiber, which is made up of aligned carbon atoms that make it easier for electrons to move through it, is to blame for the material's high conductivity. Carbon fiber is suitable for applications requiring high electrical conductivity due to its low resistivity.
- **Temperature Impacts:** The observed decline in electrical obstruction with expanding temperature can be explained by the warm development of carbon fiber. As the temperature climbs, the carbon fiber test grows, bringing about an expanded cross-sectional region.

Comparison with Conventional Testing Methods: Carbon Fiber

Carbon fiber is a material that is both light and strong. It has gained a lot of popularity in a lot of different industries, like aerospace, automotive, and sporting goods. It is an excellent choice for numerous applications due to its exceptional mechanical properties, which include corrosion resistance, low weight, and high tensile strength. However, to guarantee their performance and safety, carbon fiber products must be of the highest quality. This is where trying strategies assumes an imperative role.

The evaluation of materials' properties has traditionally been carried out using standard testing procedures. Despite their effectiveness, these strategies may not always be appropriate for carbon fiber because of its particular characteristics. In this chapter, we will compare and contrast the advantages and disadvantages of conventional and carbon fiber-specific testing techniques.

1. Tests of Tension: The mechanical properties of materials, like carbon fiber, can be determined using tensile testing. This test estimates the reaction of a material under ductile stacking, giving important data about its solidarity, versatility, and lengthening. Carbon fiber can be tested using conventional tensile machines, but they may not accurately capture its full potential.

Contrasted with ordinary strategies, specific testing machines for carbon fiber, like servo-water-driven machines, can give more exact and solid outcomes. To guarantee accurate measurements of strength, modulus, and failure modes, these machines were constructed to take into account the distinctive characteristics of carbon fiber.

2. Testing for fatigue: Weariness testing is significant for evaluating the strength and long-haul execution of materials exposed to cyclic stacking. Carbon fiber composites are often utilized in applications where weakness and obstruction are basic, like aviation parts. The material is subjected to cyclic loads during conventional fatigue testing until failure occurs.

However, due to its anisotropic nature and susceptibility to damage accumulation, carbon fiber requires special considerations during fatigue testing. To address these issues, specialized fatigue testing methods like the tension-tension fatigue test and the flexural fatigue test have been developed. With these techniques, the fatigue life of carbon fiber and its safe operating limits can be determined with greater precision.

3. Influence Testing: Carbon fiber composites are ideal for applications requiring protection from sudden loads or impacts due to their excellent impact resistance. Regular effect testing strategies, for example, the Charpy and Izod tests, are broadly utilized for different materials, but may not completely capture the way carbon fiber behaves.

Drop weight and ballistic impact tests have been developed specifically for carbon fiber composites to overcome this limitation. A more accurate representation of the material's response to dynamic loading is provided by these tests, which mimic real-world impact conditions. They assist in determining carbon fiber composites' impact resistance, damage initiation, and propagation behavior.

NDT: Non-destructive Testing

To inspect carbon fiber components without causing any damage, non-destructive testing methods are essential. Customary NDT techniques, for example, ultrasonic testing, X-beam imaging, and visual examination, have been widely utilized in different businesses. However, conventional NDT techniques face difficulties due to carbon fiber's unique properties, such as its anisotropic nature and low electrical conductivity.

Infrared thermography, acoustic emission testing, and micro-computed tomography (micro-CT) are just a few of the specialized NDT methods developed for carbon fiber. These techniques empower the discovery of deformities, delaminations, and fiber misalignments inside the carbon fiber structure, guaranteeing the quality and trustworthiness of the material.

While ordinary testing strategies have been the foundation of material assessment for quite a long time, carbon fiber's unique properties require particular testing methods. When it comes to evaluating carbon fiber composites, a comparison of conventional and specialized testing methods reveals that the latter yields result that is more accurate and dependable.

Tractable testing, weakness testing, influence testing, and non-horrendous testing, when custom-made for carbon fiber, offer improved accuracy in deciding its mechanical properties, strength, influence opposition, and imperfection identification. Manufacturers can guarantee the performance, safety, and quality of carbon fiber products in a variety of industries by utilizing these specialized testing techniques.

Applicability and Future Perspectives

A composite material known as carbon fiber is made up of fine fibers primarily composed of carbon atoms. These filaments are known for their high strength, firmness, and low weight, making them reasonable for many applications in strong mechanics. Carbon fiber is a great material for lightweight structural parts like aircraft frames, automotive parts, and sporting goods because of its unique properties.

One of the essential benefits of carbon fiber is its high strength-to-weight ratio. Carbon fiber composites can achieve comparable strength at a significantly lower weight when compared to conventional materials like steel or aluminum. Engineers can use this property to create lighter structures, use less fuel, and perform better.

The high resistance to fatigue of carbon fiber is yet another important quality. Under cyclic loading, fatigue failure typically occurs in traditional materials, resulting in structural instability and cracks. Carbon fiber composites, on the other hand, have great fatigue resistance, so they can withstand high-stress applications for a long time without losing their structural integrity.

Beyond its mechanical properties, carbon fiber can be used in many different ways. Because it is not corrosive, it is an excellent option for use in harsh environments. Additionally, carbon fiber is suitable for use in extreme temperatures due to its high thermal stability. Its suitability for the aerospace, automotive, and renewable energy sectors is made possible by these characteristics.

Carbon Fiber and Solid Mechanics in the Future

The possibilities of carbon fiber in strong mechanics are promising. Progress in assembling cycles and material exploration is supposed to additionally work on its properties, grow its scope of uses, and decrease creation costs. Key areas of advancement include:

- **Improved Manufacturing Methods:** The goal of new carbon fiber manufacturing methods is to increase quality, reduce waste, and increase production efficiency. Headways like a robotized layup, 3D printing, and high-velocity handling are being investigated to smooth out the assembling system and increase the reception of carbon fiber composites in different enterprises.
- **Design of a custom composite:** The capacity to fit the properties of carbon fiber composites as per explicit necessities is a critical area of exploration. Engineers can improve mechanical performance, reduce weight, and deal with application-specific issues by adjusting the orientation of the fibers, the resin matrix, and the architecture of the composite. The creation of individualized structures for various industries will be made easier by this strategy.
- **Recyclability and sustainability:** Sustainable carbon fiber composite manufacturing and recycling is becoming increasingly important in light of rising environmental concerns. Research is in progress to create eco-accommodating creation strategies, investigate elective unrefined substances, and layout proficient reusing cycles to limit waste and expand the life expectancy of carbon fiber parts.

Methods for the Potential Drop in Solid Mechanics

The potential drop method is a non-destructive testing method for finding damage or defects in materials and assessing their structural integrity. It works on the principle of estimating changes in electrical resistance brought about by breaks or discontinuities inside the material.

The potential drop method involves measuring the voltage drop across two points while passing a low-intensity current through the specimen. At the point when breaks or deformities are available, the ongoing stream is upset, prompting changes in opposition and a voltage drop. By breaking down these varieties, specialists can distinguish the area, size, and seriousness of imperfections, working with opportune fixes or substitutions of damaged parts.

Benefits and Applicability of the Potential Drop Method

In solid mechanics, the potential drop method has been extensively used, particularly in aerospace, nuclear power, and structural engineering. Among its advantages are:

- **Non-destructive Assessment:** The potential drop method, in contrast to destructive testing, permits the evaluation of the component's structural integrity without causing damage. This makes it an important device for reviewing basic parts and guaranteeing their protected activity.
- **Sensitivity to Minor Mistakes:** The potential drop technique can distinguish little breaks or discontinuities that may not be quickly apparent or perceivable through visual investigation. Because of this sensitivity, catastrophic failures can be detected early and avoided.
- **Cost-Effectiveness:** A low-cost method for assessing structural integrity and identifying flaws is the potential drop method. It wipes out the requirement for costly and tedious reviews or broad dismantling of parts, saving both time and assets.

The Potential Drop Method's Future Prospects

Proceeding with research and mechanical headway is probably going to improve the potential drop strategy's abilities and extend its materiality in strong mechanics. Some perspectives for the future include:

- **Incorporation with Advanced Innovations:** The potential drop method can be combined with digital technologies like artificial intelligence and machine learning algorithms to automate the analysis process and improve the accuracy of defect detection. The performance and safety of structural components would be enhanced by this integration, which would make it possible to perform predictive and real-time monitoring.

- **Multi-Scale Examination:** The potential drop method may be able to evaluate materials at various length scales thanks to advancements in imaging and data processing. This would make it easier to find defects at the micro- and nanoscales, which would make it possible to use them in new areas like nanotechnology and advanced materials.
- **Construction of portable systems:** Portable and handheld drop systems are currently being developed. These minimal gadgets would consider nearby reviews, diminishing margin time, and further developing upkeep effectiveness in different enterprises, including development, transportation, and framework.

The potential drop method and carbon fiber's use in solid mechanics have enormous potential to advance material science and structural engineering. Carbon fiber composites offer outstanding mechanical properties, and progressing innovative work endeavors are supposed to further improve their presentation, extend their applications, and lessen creation costs. The potential drop method is a useful non-destructive testing method that allows for precise defect detection and structural integrity evaluation simultaneously. The future points of view of both carbon fiber and the potential drop technique demonstrate a way towards maintainable and inventive arrangements, adding to more secure and proficient designs across different businesses.

CONCLUSION

As of late, carbon fiber has emerged as a progressive material with gigantic potential in different enterprises, especially in the field of strong mechanics. It is favored for a wide range of applications, including sports equipment and construction, as a result of its exceptional strength, low weight, and remarkable mechanical properties. Closely related to carbon fiber, the potential drop strategy has proven to be a priceless procedure for assessing the underlying trustworthiness of materials and parts. In this conclusion, we will discuss the significance of carbon fiber and the potential drop method in solid mechanics, as well as a summary of the main findings.

Due primarily to their exceptional strength-to-weight ratio, carbon fiber composites have revolutionized materials engineering. Carbon fibers are embedded in an epoxy resin-like matrix that provides additional structural support for these composites. Carbon fiber is an excellent choice for applications where mechanical performance and weight reduction are crucial due to its unique combination of high strength, stiffness, and low weight. Carbon fiber composites are gaining popularity in a variety of industries, including the automotive industry, where improved performance and safety are sought, and the aerospace industry, where fuel efficiency is of the utmost importance.

The high tensile strength of carbon fiber composites, which is higher than that of traditional materials like steel or aluminum, is one of the most important benefits. Because of this property, lightweight structures can be made without sacrificing strength, which improves performance and fuel economy. Additionally, carbon fiber composites have excellent fatigue resistance, making them ideal for cyclic loading applications like wind turbine blades or aircraft wings.

Moreover, carbon fiber composites offer brilliant consumption resistance, which is a critical benefit over metals like steel or aluminum that are inclined to erosion. Carbon fiber composites are an excellent choice for applications in harsh environments, such as the chemical or marine industries, due to this property. Also, the non-attractive nature of carbon fiber makes it profoundly alluring for applications where electromagnetic impedance should be limited, like in aviation or clinical gear.

The potential drop method has proven to be an essential tool for assessing the structural integrity of materials and components in solid mechanics. To determine whether a material has cracks or defects, this non-destructive testing method uses the measurement of electrical potential differences. The potential drop method is based on the idea that a crack or defect causes a localized change in electrical resistance by acting as a discontinuity in the material. It is possible to determine the size and location of the defect by passing a low-level direct current through the material and measuring the potential difference across the crack.

The potential drop strategy offers a few benefits over customary examination methods. First and foremost, it is non-destructive, making it possible to evaluate structures without compromising their integrity or causing damage. This feature is especially useful in industries like aerospace, where components are expensive and failure has severe repercussions. Additionally, the potential drop method has high sensitivity, making it possible to identify even the tiniest flaws or cracks that may not be apparent to the naked eye. This capability reduces the likelihood of a catastrophic failure by enabling early detection and preventative maintenance.

For a variety of industries, the potential drop method and carbon fiber combine greatly. Manufacturers can guarantee carbon fiber composites' quality and structural integrity throughout their product's lifespan by employing the potential drop method. This is especially important because carbon fiber composites can be damaged by impacts or too much weight, which can cause hidden cracks or defects. Such flaws can be detected and monitored using the potential drop method, allowing for prompt replacements or repairs to prevent catastrophic failures.

3

Plastic Deformation and Periodic Freezing of a Solid Mechanics

ABSTRACT

Plastic twisting is a principal peculiarity in strong mechanics that happens when a material goes through long-lasting shape changes because of the utilization of outside powers. Understanding the mechanics of plastic disfigurement is pivotal for different modern applications and designing plans. This abstract presents a thorough report on the impact of occasional freezing on plastic twisting in solids.

The review starts by giving an outline of plastic twisting and its basic systems. It investigates the idea of the disengagement hypothesis, which makes sense of the development of precious stone deformities called separations and their part in versatility. The job of pressure and endless strain rate in plastic deformity is additionally talked about, featuring the significance of these elements in deciding material conduct under outer stacking.

The ensuing segment digs into the peculiarity of intermittent freezing and its effect on the plastic deformity. Occasional freezing involves exposing a strong material to cyclic variations in temperature, bringing about the development of ice gems inside the material. The association between the developing ice gems and the strong network presents extra burdens and strains, prompting changes in the plastic deformity's behavior.

The review explores the impact of freezing cycles on the mechanical properties of materials, including their yield strength, flexibility, and sturdiness. Trial procedures like malleable testing and microstructural examination are utilized to evaluate the progressions initiated by intermittent freezing. The outcomes reveal that freezing cycles can change the separation thickness, grain size, and stage dissemination inside the material, thus influencing its mechanical reaction.

Moreover, mathematical reenactments are directed at acquiring further bits of knowledge about the plastic deformity's behavior under occasional freezing conditions. Limited component examination and atomic element recreations are utilized to display the mind-boggling connections between separations, precious stone imperfections, and ice gems. These reenactments give a definite comprehension of the basic systems overseeing plastic disfigurement and the impact of freezing cycles.

The discoveries of this study have critical ramifications for a scope of uses, including cryogenic designing, chilly climate materials handling, and designs exposed to occasional freezing conditions. By clarifying the systems of plastic deformity under freezing cycles, architects and analysts can foster stronger materials and further develop plans for different businesses, like aviation, energy, and transportation.

Keywords: *plastic twisting, strong mechanics, disengagement hypothesis, occasional freezing, mechanical properties, freezing cycles, yield strength, pliability, sturdiness, exploratory procedures, mathematical reproductions, cryogenic designing, materials handling, the underlying model*

INTRODUCTION

A fundamental idea in solid mechanics, which studies how materials behave when loads are applied, is plastic deformation. It alludes to the long-lasting change in shape and size of a material when exposed to outside forces past its flexible breaking point. Dislocations in the material's crystal structure move, allowing for the rearrangement of atoms or molecules, which results in plastic deformation.

The design of structural components for the production of everyday goods requires an understanding of plastic deformation in engineering. Materials' mechanical properties, such as toughness, ductility, and strength, are significantly influenced by it. Accordingly, broad exploration has been conducted to research the systems and variables impacting plastic distortion.

The periodic freezing effect that is observed in certain materials is one interesting phenomenon that is connected to plastic deformation. The intermittent halting or slowing down of plastic deformation during cyclic loading or deformation processes is referred to as periodic freezing. This conductivity has been seen in a few materials, including metals, polymers, and ceramics, and has drawn huge consideration from scientists.

The mechanical properties and behavior of materials are profoundly affected by the periodic freezing effect, according to research. It can influence the exhaustion life of parts, the arrangement of breaks and deformities, and the general presentation of designs exposed to cyclic stacking. In engineering design and structural integrity evaluation, it is, therefore, crucial to comprehend the underlying mechanisms of periodic freezing and its interaction with plastic deformation.

In solid mechanics, the phenomena of plastic deformation and periodic freezing are the subject of this study. The purpose of this chapter is to provide a comprehensive understanding of the mechanisms underlying plastic deformation and the causes of periodic freezing. In addition, the chapter will investigate the effects of periodic freezing on structural behavior and the mechanical properties of materials.

To accomplish these targets, a thorough writing survey will be directed to assemble existing information on the plastic deformity and occasional freezing. The theoretical models, experimental studies, and computational simulations that have been carried out in this area will be covered in the review. This chapter aims to present a coherent and up-to-date comprehension of plastic deformation and periodic freezing by analyzing and synthesizing existing research.

Moreover, this chapter will feature the reasonable ramifications of the occasional freezing impact in designing applications. It will talk about how periodic freezing can be used to increase the fatigue life of materials and make structures that are hit by cyclic loads last longer. The challenges and limitations of detecting and predicting periodic freezing will also be discussed in the chapter, as will potential directions for future research in this field.

The phenomena of periodic freezing and plastic deformation in solid mechanics are intriguing and have significant repercussions for the behavior of materials and the performance of structures. This exploratory chapter means to give an exhaustive outline of plastic twisting and occasional freezing, including their hidden components, consequences for mechanical properties, and useful ramifications. By investigating the present status of information and recognizing research holes, this chapter will contribute to propelling comprehension as we interpret plastic deformity and occasional freezing and cultivate further exploration in this significant field of study.

Background

Plastic Twisting

The plastic deformity is portrayed by the extremely durable change in shape or size of a material when exposed to outside forces. Plastic deformation is irreversible, in contrast to elastic deformation, in which the material returns to its original shape after the stress has been removed. At the point when a material encounters plastic twisting, its inner construction goes through revisions at the nuclear or minuscule level, bringing about separations and different deformities.

When the applied stress is greater than the material's yield strength, plastic deformation occurs. The stress at which a material's behavior changes from elastic to plastic is represented by its yield strength. Past the yield point, the material begins to display strain and solidify, where the protection from deformity expands because of the arrangement of new separations and the connection between existing ones.

Plastic deformation is caused by slip, twinning, and dislocation motion, among other mechanisms. When dislocations move along particular crystallographic planes, one part of the crystal slides past another. This is called a slip. Twinning, then again, includes the development of an identical representation of a piece of the precious stone design, bringing about an adjustment of shape. Plastic flow is made possible by dislocations moving through the crystal lattice, which is referred to as dislocation motion.

The deformation of plastic has numerous practical repercussions. In primary design, it is urgent to consider the plastic behavior of materials to guarantee the security and dependability of designs under various stacking conditions. Versatility likewise impacts material handling tasks like manufacturing, rolling, and expulsion, where molding and shaping tasks depend on the capacity of materials to go through plastic twisting without disappointment.

The repeated application of cyclic stresses to a material is referred to as periodic freezing, also known as cyclic loading or fatigue loading. This peculiarity usually happens in mechanical frameworks exposed to fluctuating burdens, for example, turning apparatus, scaffolds, and airplane parts. The redundant idea of cyclic stacking can prompt moderate harm and disappointment in materials, in any event, when the singular burdens are underneath the material's definitive strength.

Due to its cumulative effect on materials, cyclic loading presents unique challenges in solid mechanics. Each pressure cycle causes infinitesimal harm, like separation development, break commencement, and grain limit sliding. These small damages build up over time and cause larger failures, most often fatigue cracks.

The magnitude and frequency of applied stresses, the strength and ductility of the material, the presence of stress concentrations, and environmental conditions all have an impact on a material's fatigue life. Designing structures and components that can withstand cyclic loading without premature failure necessitates an understanding of fatigue behavior.

Materials' fatigue life is frequently predicted using fatigue analysis methods like stress-life (S-N) and strain-life (-N) approaches. These strategies include exploratory testing and measurable investigation to lay out weariness bend and foster weakness plan rules for explicit applications.

Objectives

Plastic twisting alludes to the long-lasting change in the state of a strong material under the activity of outside powers, while occasional freezing includes the reversible change of a strong material between various stages because of temperature changes. The purpose of this chapter is to discuss the goals of studying periodic freezing and plastic deformation, as well as their implications for solid mechanics.

Understanding the Deformation of Plastic

Understanding how materials react to external loads is the primary goal of studying plastic deformation. In designing structures and anticipating their mechanical responses, this knowledge is essential. The following are some specific goals of studying plastic deformation:

- **Portraying Material Reaction:** Exploring the material reaction under various stacking conditions decides its solidarity, flexibility, sturdiness, and other mechanical properties. The design of structures that can withstand anticipated loads and the selection of appropriate materials for particular applications rely heavily on this information.
- **Plasticity Modeling:** Creating numerical models and constitutive conditions that precisely portray the plastic way of behaving materials is another goal. The deformation and failure mechanisms of structures under various loading scenarios can be predicted with the assistance of these models.
- **Dissecting Plastic Deformity Systems:** It is essential to comprehend the microscopic mechanisms that underlie plastic deformation, such as dislocation motion, twinning, and interactions at grain boundaries. The material's plastic flow, strain hardening, and deformation localization can all be better understood through an examination of these mechanisms, which aids in the creation of cutting-edge materials and processing methods.

Examining Frequent Freezing

Phase transformations and structural changes caused by temperature variations in materials are examined in the study of periodic freezing. The targets of researching intermittent freezing include:

- **Stage Outline Assurance:** Building stage graphs for materials empowers the recognizable proof of different stages and their soundness ranges at various temperatures and tensions. These outlines give vital data to material determination, handling, and understanding of the microstructural changes during stage changes.

- **Warm Pressure Examination:** Intermittent freezing prompts warm development and compression, causing warm burdens inside a material. Examining these burdens assists architects and researchers with planning materials and designs fit for enduring warm cycling without disappointment.
- **Amalgam Plan and Intensity Therapy:** The study of periodic freezing contributes to the development and improvement of alloy compositions and heat treatment procedures. Understanding what different cooling rates and warm cycles mean for the material's microstructure empowers the fitting of mechanical, electrical, and warm properties as indicated by unambiguous application prerequisites.

Synergistic Impacts of Plastic Deformity and Occasional Freezing

Additional research goals arise from the interaction between plastic deformation and periodic freezing. Some important goals include:

- **Phase Changes Caused by Strain:** It is essential to investigate how materials' phase transformations can be induced or influenced by plastic deformation. The development of materials with improved mechanical properties and the optimization of processing methods are aided by an understanding of these strain-induced phase transformations.
- **Analysis of Remaining Stress:** Residual stresses can form within a material when plastic deformation and periodic freezing are combined. When determining the material's stability, fatigue resistance, and susceptibility to stress corrosion cracking, it is essential to analyze these residual stresses.
- **Execution Forecast:** The long-term performance of structures and components is impacted by the synergistic effects of plastic deformation and periodic freezing. Creating prescient models that consider these impacts helps with assessing the assistance life and sturdiness of materials under sensible working circumstances.

Scope and Significance

Plastic deformation and occasional freezing are urgent peculiarities in the field of strong mechanics. Under the influence of external forces or changes in temperature, these processes involve the transformation of a solid material's structure and shape. For engineering, materials science, and manufacturing, understanding the scope and significance of plastic deformation and periodic freezing is essential. This chapter plans to investigate the extension and meaning of these peculiarities, their applications, and their effect on material ways of behaving and execution.

Plastic deformation and periodic freezing's scope:

- **Material ways of behaving:** A material's permanent change in shape or size beyond its elastic limit is known as plastic deformation. It occurs when the material's yield strength is exceeded by stress. Plastic twisting can prompt different material ways of behaving, for example, strain solidifying, flexibility, and work solidifying. Investigating the stress-strain relationship, mechanical properties, and failure mechanisms of materials under various loading conditions are all necessary for comprehending the scope of plastic deformation.
- **Primary Examination:** Structural analysis and design rely heavily on plastic deformation. It influences factors like load distribution, stress concentration, and failure modes, influencing the behavior of structural components under load. Architects and creators should consider plastic deformation to guarantee the honesty and wellbeing of designs, like structures, spans, and mechanical parts. Structural engineers can improve structural performance, reduce material consumption, and optimize designs by accurately predicting and controlling plastic deformation.
- **Methods of Production:** Plastic deformation plays a significant role in several manufacturing processes. Plastic deformation is used to shape metals into the forms that are desired during metal forming operations like forging, rolling, and extrusion. Plastic disfigurement additionally assumes an essential role in other assembling strategies like infusion forming, blow embellishment, and sheet metal bowing. By controlling plastic twisting during assembling, designers can accomplish exact aspects, desired mechanical properties, and further developed item quality.
- **Material Handling:** The extent of plastic deformation extends to material handling strategies like virus work and intensity therapy. In cold working, plastic deformation is applied at low temperatures to enhance the strength and hardness of materials' mechanical properties. Heat treatment, on the other hand, alters materials' microstructure and properties by combining controlled heating and cooling. Understanding the extent of plastic deformation in material handling empowers the improvement of cutting-edge materials with custom-made properties for explicit applications.

The Importance of Periodic Freezing and Plastic Deformation

- **Mechanical Plan and Underlying Honesty:** The impact that plastic deformation has on mechanical design and structural integrity is what makes it important. Engineers can design structures and mechanical components with appropriate safety factors, such as yield strength, ductility, and fatigue resistance, by comprehending the behavior of materials under load. Exact expectations and control of plastic disfigurement are essential for guaranteeing the unwavering quality and life span of designed frameworks.

- **Material Choice and Advancement:** Material selection and optimization are significantly influenced by plastic deformation. Engineers can select materials that best meet a given application's requirements by taking factors like yield strength, ductility, and formability into account. The ability to anticipate and control plastic deformation enables the optimization of material processing routes, enhancing material properties, and lowering production costs.
- **Prevention and Analysis of Failure:** Understanding plastic deformation is fundamental for disappointment investigation and counteraction. Engineers can identify potential failure modes and take preventative measures by studying the mechanisms of plastic deformation and how they interact with other factors like stress concentrations and environmental conditions. This information is essential in businesses where disappointment can have extreme results, like aviation, cars, and oil and gas.
- **Materials Science and Engineering Developments:** Plastic deformity research adds to advances in materials science and design. Scientists and engineers can make new materials with better properties, better performance, and better resistance to plastic deformation by studying the fundamental aspects of plastic deformation like dislocation behavior, grain boundaries, and phase transformations. Aerospace, electronics, renewable energy, and biomedical engineering are just a few of the many fields where these advancements can be used.

PLASTIC DEFORMATION

Definition and Overview

A fundamental idea in materials science and engineering is plastic deformation, which describes the material's permanent change in shape or size when it is subjected to external forces or loads. Plastic deformation involves the movement and rearrangement of atoms or molecules within the structure of the material, in contrast to elastic deformation, which is reversible and does not result in permanent changes. This peculiarity assumes a pivotal role in different fields, including fabricating, designing plans, and metallurgy. The underlying mechanisms, key parameters, and practical implications of plastic deformation are examined in this chapter, which provides a comprehensive definition and overview.

The deformation of plastic is defined as:

The irreversible deformation of a material when it is subjected to external forces or stresses beyond its elastic limit is known as plastic deformation. It involves permanent atom or molecular displacement within the crystal lattice or amorphous structure of the material, resulting in changes in shape, size, and mechanical properties. Depending on the applied stress conditions and the material's response, plastic deformation can manifest as bending, stretching, compressing, or twisting.

Components of Plastic Distortion

The crystal structure, temperature, strain rate, and other properties of the material all play a role in how plastic deformation occurs. The most typical mechanisms are:

- **Slip:** Slip is the prevailing system in glasslike materials like metals and composites. Dislocations, or line defects in the crystal structure, are moved through crystal planes in this process. At the point when pressure is applied, separations coast along favored crystallographic planes, causing plastic disfigurement.
- **Twinning:** Another mechanism that has been observed in some crystalline materials is twinning. It includes the development of identical representation spaces, called twins, inside the precious stone cross-section. Twinning can happen because of explicit pressure conditions and is responsible for plastic distortion in materials like magnesium and titanium.
- **Viscous Motion:** Through a process known as viscous flow, amorphous materials like polymers and glasses exhibit plastic deformation. Amorphous materials lack long-range order, and their atoms or molecules move more randomly than in crystalline materials. Under pressure, these materials go through nuclear modifications, bringing about long-lasting twisting.

Factors that Affect Plastic Deformation

A few boundaries impact the degree and nature of plastic twisting in materials:

- **Stress:** The type and extent of plastic deformation are determined by the applied stress's magnitude and direction. Ductile, compressive, shear, or torsional stresses can cause various types of plastic disfigurement.

- **Stress Level:** The rate at which distortion happens, known as the strain rate, influences the way materials behave during plastic deformity. While slower strain rates can improve ductility, higher strain rates frequently result in behavior that is more brittle or prone to fracture.
- **Temperature:** Plastic deformation is significantly influenced by temperature. The flow of plastic is made easier by increased atomic mobility at higher temperatures. Notwithstanding, outrageous temperatures can likewise initiate stage changes or warm relaxation, adjusting the material's deformity behavior.
- **Merits of the Material:** Plastic deformation is significantly influenced by inherent material properties like grain size, crystal structure, composition, and defects. Materials with more modest grain sizes or more elevated levels of imperfections will generally show expanded strength and further developed protection from the plastic deformity.

Implications for the Real World from Plastic Deformation

The deformation of plastic has numerous practical repercussions across a variety of industries.

- **Manufacturing:** Forging, rolling, extrusion, and wire drawing are just a few of the manufacturing processes where plastic deformation is frequently used. By causing controlled plastic deformation, these processes reshape and shape materials into the desired shapes and dimensions.
- **Design of the Structure:** To create structures that are both secure and dependable, it is essential to comprehend the plastic deformation behavior of materials. Engineers must take yield strength, strain hardening, and ductility into account to ensure that structures can withstand anticipated loads without breaking down completely.
- **Material Choice:** The resistance of various materials to plastic deformation varies. This information assists engineers with choosing fitting materials for explicit applications. For load-bearing components, for instance, materials with high strength but low ductility are ideal, while applications requiring greater deformation capacity employ materials with greater ductility.

Mechanisms of Plastic Deformation

The key to designing and optimizing materials for specific applications is comprehending the mechanisms that underlie plastic deformation. The mechanisms of plastic deformation, including dislocation motion, slip systems, and deformation mechanisms in various materials, are examined in this chapter.

Displacement Movement

Line defects or irregularities in materials' crystal lattice structures are called dislocations. Under external stress, they can move through the lattice, causing plastic deformation. Disengagement movement happens through two essential components: edge separation and screw disengagement.

- **Edge Dislocation:** When an additional half-plane of atoms is added to the crystal lattice, it causes an edge dislocation. Atomic bonds are repeatedly broken and reformed during an edge dislocation's movement. This movement brings about the separation line moving in line with the applied pressure, causing a shear deformation.
- **Negative Dislocation:** When the crystal lattice is twisted around the dislocation line, a screw dislocation occurs. The movement of a screw disengagement includes the separation line moving in line with the applied pressure through a twisting-like movement. This results in a shear deformity-like edge separation.

Slip Frameworks

The most prevalent method of plastic deformation in crystalline materials is slip. Slip happens when disengagements move along unambiguous planes, known as slip planes, and in unambiguous headings, known as slip bearings. A slip system is the combination of a slip plane and a slip direction.

Because they require less energy, specific slip systems in each crystal structure are better suited for dislocation motion. 111 slip planes and 110> slip directions, for instance, are common in face-centered cubic (FCC) metals like gold, copper, and aluminum. The majority of body-centered cubic (BCC) metals, such as iron and tungsten, have slip directions of 111> and 110>.

The movement of separations along slip planes produces plastic deformation as adjoining nuclear planes slide past one another. The material undergoes macroscopic deformation as a result of this movement, which results in shear strain.

Mechanisms for Deformation in Various Materials

Metals, ceramics, and polymers are just a few examples of materials where the mechanisms of plastic deformation can differ.

- **Metals:** Plastic deformation is mostly caused by dislocations moving in metallic materials. Because of the somewhat high malleability of metals, separations can undoubtedly occur through the precious stone cross-section, taking into account critical plastic disfigurement. Mechanisms like dislocation multiplication, cross-slip, and climb facilitate dislocation motion.

- **Ceramics:** In comparison to metals, the response of ceramic materials to plastic deformation is more complex. Since pottery has a high ionic or covalent bonding character, disengagement movement is restricted. Instead, transformation-induced plasticity, grain boundary sliding, and twinning dominate ceramic deformation.
- **Polymers:** In polymers, the plastic deformity is essentially brought about by the sliding and unraveling of polymer chains. Polymers are amorphous, allowing for chain movement and reorganization, which leads to plastic flow. Polymers can also respond to stress in a time-dependent manner and exhibit viscoelastic behavior.

Plasticity and hardening of the strain

Work hardening, also known as strain hardening, is a phenomenon that occurs during plastic deformation in materials. Dislocations interact with one another as they move and accumulate, preventing further dislocation motion. Strain hardening occurs when the material's resistance to deformation rises as a result of this interaction. Plastically deformed materials exhibit increased strength and decreased ductility due to strain hardening.

On the other hand, a material's plasticity is its capacity for plastic deformation. It is a desirable quality for materials that are used in situations where plastic flow or shape changes are required. The components examined before, for example, separation movement and slip, add to the general versatility of materials.

In materials science and engineering, plastic deformation is a crucial process that enables the shaping, forming, and production of various products. The comprehension of the fundamental instruments of plastic disfigurement, including separation movement, slip frameworks, and material-explicit twisting systems, is fundamental for planning and improving materials for explicit applications. By acquiring experience in these systems, specialists, and researchers can foster new materials with further developed properties and execution in different enterprises, from aviation and cars to hardware and biomedical applications.

Stress-Strain Relationships

Understanding the relationships between stress and strain, as well as the phenomenon of plastic deformation, is an essential component of this research. The fundamental concepts of stress and strain are utilized to describe the behavior of materials under external forces. The permanent change in size or shape that occurs when a material goes beyond its elastic limit is known as plastic deformation. We will investigate the mechanisms of plastic deformation in materials and the relationships between stress and strain in this chapter.

Anxiety: The force exerted on a material per unit area is called stress. It addresses the inner obstruction of a material due to deformation brought about by outside powers. The cross-sectional area over which the force acts is divided by the applied force to determine the stress. Pascal (Pa), which is equivalent to one Newton per square meter (N/m2), is the SI unit for stress.

In contrast, strain measures the material's deformation concerning its original size or shape. It is calculated by dividing the original length (or dimension) by the change in length. Strain is a dimensionless amount, yet it is often communicated as a rate or in microstrains.

The curve of stress and strain: The pressure-strain relationship for a material is normally addressed by a pressure-strain bend. This curve provides useful information about the material's mechanical properties by demonstrating how it responds to stress. The pressure strain bend can be separated into a few unmistakable locales:

- **Versatile District:** Around here, the material acts flexibly, implying that it gets back to its unique shape and size after the applied powers are removed. The material adheres to Hooke's Law, which states that stress is directly proportional to strain within the elastic limit and that the relationship between stress and strain is linear.
- **Yield Point:** Some materials have a yield point, or stress level, at which they begin to undergo plastic deformation beyond the elastic region. The material experiences permanent deformation at the yield point, and the stress-strain curve departs from linearity.
- **Region of plastic:** In the plastic locale, the material goes through critical, long-lasting twisting with expanding pressure. The pressure strain bend shows strain solidifying, where the material becomes more grounded and more impervious to deformation as plastic disfigurement advances.
- **Necking and Break:** In the long run, the material might encounter necking, which is restricted distortion that brings about a decrease in the cross-sectional region. Following necking is a fracture, in which the material fails and breaks under stress.

Components of Plastic Distortion: The movement of dislocations within a material's crystal structure causes plastic deformation.

Separations are line absconds that address locale of cross-sectional bending or misalignment. At the point when stress is applied to a material, separations can move, prompting a plastic stream and super-durable deformation.

There are a few components through which disengagements move:

- **Slip:** Slip is the most well-known instrument of disengagement development. It occurs when dislocations respond to applied shear stress by moving along particular crystallographic planes. Slip is the cause of the plastic deformation seen in metals and can occur in both crystalline and polycrystalline materials.
- **Twinning:** Twinning is another way that plastic deformation occurs, and it is particularly evident in some crystal structures. The movement of twin boundaries results in the formation of mirror-image regions within a crystal during twinning. This component can contribute to plastic deformity in materials like magnesium and titanium.
- **Diffusion:** Dissemination-controlled creep is a plastic deformation instrument that includes the development of particles through the cross-section of material. When subjected to constant stress and high temperatures, it is significant. Dissemination permits molecules to rework, prompting separation movement and a plastic stream.
- **Climb in dislocation:** Disengagement climb happens when separations move opposite to their slip planes through the ingestion or outflow of point absconds. This component is seen in materials with a high centralization of openings or interstitials.

Plasticity Models

Versatility models give significant bits of knowledge into the way materials behave under various stacking conditions and are widely utilized in the planning and examination of designs. In this chapter, we will investigate pliancy models and their applications in understanding plastic deformation processes.

Understanding the difference between elasticity and plasticity is essential before delving into models of plasticity. The reversible change in shape that occurs when a material is subjected to external forces and returns to its original shape when the forces are removed is referred to as elastic deformation. Plastic deformation, on the other hand, results in shape changes that last forever and do not completely restore the material to its original state.

Plasticity Schematics

The elastic-perfectly plastic model, which assumes linear elasticity up to yield stress and then a sudden transition to plastic deformation, is the simplest plasticity model. The material in this model does not harden under strain, so the stress stays the same after yielding. The flexible, completely plastic model is numerically advantageous, but it distorts the real way materials behave, especially in complex stacking conditions.

Strain-hardening models Strain-hardening models were created to overcome the limitations of the elastic-perfectly plastic model. The material's capacity to undergo additional plastic deformation beyond yielding is taken into account in these models, which increases the stress required for further deformation. The isotropic hardening model is the strain hardening model that is utilized the most frequently. In this model, the yield stress increases with plastic strain but remains isotropic (i.e., unaffected by the orientation of the material).

Kinematic hardening models account for the evolution of the yield stress with plastic strain and loading history, in contrast to isotropic hardening models, which assume constant yield stress in all directions. The material's memory of previous deformation and its effect on subsequent loading is taken into consideration in these models. Although kinematic hardening models are more sophisticated, calibration necessitates more intricate mathematical formulations and experimental data.

Combinational hardening models In some instances, it is necessary to take into account both kinematic and isotropic hardening mechanisms. By incorporating both effects, combined hardening models combine the advantages of isotropic and kinematic hardening. Although they require additional parameters and experimental data for calibration, these models provide a more accurate representation of material behavior under complex loading conditions.

Yield Measures

Yield measures characterize the circumstances under which a material changes from versatile to plastic. Regularly utilized yield rules incorporate the von Mises model, Tresca basis, and Mohr-Coulomb standard. A good representation of the behavior of ductile materials is provided by the von Mises criterion, also known as the maximum distortion energy criterion.

Plasticity Models and the Finite Element Method The finite element method (FEM) frequently makes use of plasticity models to simulate and analyze the behavior of structures that are subjected to a variety of loading conditions. Plasticity models are applied to each discrete element of the structure using FEM to predict and simulate the plastic deformation response. Engineers can optimize designs and predict failure under a variety of loading scenarios by combining FEM and plasticity models.

Applications of Plasticity Models Plasticity models are used in a lot of different areas, like manufacturing, civil engineering, aerospace, automotive, and aerospace. Predicting the behavior of materials during rolling, forging, and extrusion processes is made easier by these models. They are also essential for designing safe and effective structures and evaluating the structural integrity of components under various loading conditions.

PERIODIC FREEZING IN SOLID MECHANICS

Concept and Definition

The term "periodic freezing" refers to a cyclical or periodic behavior that can be found in some materials. This behavior is characterized by a brief loss of mechanical functionality as a result of the formation of ice crystals. The concept and definition of periodic freezing in solid mechanics will be examined in this chapter, along with its underlying mechanisms, effects, and potential applications.

Definition

Occasional freezing, otherwise called cyclic freezing, is a peculiarity experienced in unambiguous materials where dull freezing and defrosting cycles bring about the development and resulting liquefying of ice gems. It typically occurs in biological tissues, porous materials like soil, concrete, asphalt, or composites that contain water or moisture. The material's water expands as it solidifies into ice during each cycle of freezing, putting pressure on the surrounding solid matrix. After defrosting, the ice dissolves, and the tension feels better. The material's structural integrity, properties, and performance can all be altered as a result of this cyclical process of freezing and thawing, which can have significant mechanical effects.

How Periodic Freezing Works

The interaction of two fundamental processes is primarily responsible for the occurrence of periodic freezing: the mechanical response of the solid matrix and the water's phase shift.

Water's morphology: Water is unique because of its capacity to expand when it freezes. Ice crystals are formed when water freezes, moving from a liquid to a solid state. During this change, the water particles organize themselves into a hexagonal cross-section, bringing about an expansion in volume. In permeable materials or composites, the presence of water permits the entrance of the fluid into the pore spaces. When water freezes, its expansion presses on the solid matrix around it, causing stress and deformation.

The Solid Matrix's Response to Mechanical Forces: The strong network in materials encountering occasional freezing assumes a pivotal role in deciding the by and large mechanical way of behaving. Contact forces and constraints interact with the expanding ice crystals. The stiffness, porosity, permeability, cohesion, and other properties of the material influence how the solid matrix responds to these forces. The solid matrix experiences repeated loading and unloading during cyclic freezing and thawing, resulting in cumulative deformation, cracking, or damage. The microstructure, composition, and moisture content of the material all have an impact on these mechanical effects.

Effects and Impacts

The mechanical behavior and properties of materials can be significantly altered by periodic freezing. Some notable outcomes include:

- **Injuries and Decline:** The material's repeated expansion and contraction during cycles of freezing and thawing can result in cracking, spalling, or disintegration due to mechanical stresses that are greater than its strength. The accumulation of harm over the long run compromises the underlying uprightness and toughness of the material, presenting potential dangers. This impact is especially obvious in framework materials like substantial asphalt, scaffolds, or building establishments.
- **Penetrability and Transport Properties:** Occasional freezing impacts the penetrability and transport properties of materials. The ability of a material to retain or transmit heat, gases, or fluids can be altered by the formation of cracks or fractures during freezing cycles. This can affect the exhibition of materials in applications where controlling the liquid stream, protection, or intensity of movement is basic.
- **Environmental and Biological Systems:** In natural frameworks, occasional freezing can influence the reasonability and usefulness of living tissues or creatures. In plant tissues, for instance, cycles of freezing and thawing can cause damage to cells by disrupting cell membranes. The availability of nutrients, root growth, and the dynamics of the ecosystem as a whole can all be affected by soil freezing and thawing cycles in cold climates.
- **Applications and Moderation Systems:** Although periodic freezing is frequently regarded as a negative occurrence, it can also

be harnessed or mitigated with the right strategies. The following are important applications and mitigation strategies:

1. **Material Science**: Understanding the impacts of intermittent freezing on materials takes into account the advancement of further developed design arrangements. Engineers can improve materials' resistance to freezing-induced damage and extend their service life by incorporating additives, modifying pore structures, or optimizing their composition.
2. **Geotechnical Design:** In geotechnical design, understanding occasional freezing is essential for planning establishments, holding walls, or underground designs in cool locales. Legitimate protection, seepage frameworks, or ground support strategies can be utilized to limit the inconvenient impacts of freezing and defrosting on the steadiness and execution of designs.
3. **Cryogenic Storage:** In the cryopreservation of biological samples, tissues, or cells, periodic freezing has been utilized in fields like medicine and biotechnology. Cryoprotectants and controlled freezing and thawing procedures make it possible to preserve biological material for clinical, transplant, or research purposes.

Causes and Effects

Predicting material behavior, mitigating potential risks, and designing resilient structures all depend on understanding the causes and effects of periodic freezing. The primary causes of periodic freezing and their effects on various aspects of solid mechanics are examined in this chapter.

I. The Reasons for Frequent Freezing:

- **Temperature Varieties:** Temperature variances below the edge of freezing are the key reason for occasional freezing. Diurnal and occasional changes in temperature open materials to substitute freezing and defrosting cycles. Climate, location, altitude, and the absence of adequate insulation can all have an impact on these temperature variations.
- **Content of moisture:** The frequency of periodic freezing is greatly influenced by the amount of moisture present. Particularly vulnerable are materials with high water content, such as porous rocks, saturated soils, and construction materials like concrete. The material's water molecules expand when it freezes, causing internal stresses that cause cracking, disintegration, and then thawing when the temperature rises.

- **Permeability and porosity:** The presence of interconnected voids, pores, or cracks inside a material influences its capacity to hold and convey dampness. Exceptionally permeable and porous materials work with the invasion of water, which, after freezing, can cause development and harm due to ice arrangement. Grain size, compaction, and material composition all influence the degree of porosity and permeability.
- **Cycling in the Freeze-Thaw Situation:** Rehashed freeze-defrost cycles fuel the harm brought about by occasional freezing. Each freezing cycle adds to the gathering of inside anxieties, prompting microcracks and primary debasement. Water expands deeper into the material with each subsequent thawing cycle, widening existing cracks and opening up new ones, thereby speeding up the overall deterioration process.

II. Consequences of Repeated Freezing:

- **Mechanical Strength Decrease:** The continued freezing and defrosting of materials, step by step, debilitates their mechanical properties. As ice extends inside the material, it incites inner burdens that surpass the strength of the network. This reduces the structure's load-bearing capacity, stiffness, and overall integrity due to cracking, spalling, and loss of cohesion.
- **Changes in Dimensions:** Solid materials may experience dimensional shifts as a result of periodic freezing. Deformations, warping, and surface irregularities can result from the ice's expansion and contraction during thawing. Engineering structures can suffer from misalignments, joint failures, and other functional issues as a result of these alterations in shape and dimensions.
- **Strength and Administration Life Decrease:** Periodic freezing has a significant negative impact on the durability and service life of materials and structures. They become more susceptible to other environmental factors like chemical reactions, abrasion, and corrosion as a result of the continual freeze-thaw cycles. This can prompt untimely disappointment, requiring successive fixes, substitutions, and expanded upkeep costs.
- **Effects on the environment:** Natural environments as well as human-made structures are affected by periodic freezing. Cyclic freezing has the potential to disrupt the structure of ecological systems, affecting the availability of nutrients, root growth, and water infiltration. It can likewise prompt the ice hurl peculiarity, which can harm vegetation, upset environments, and influence foundations based on frozen ground, like streets, pipelines, and structures in cool districts.

Experimental Observations

The goal of this study is to provide a comprehensive analysis of experimental findings on periodic freezing. By looking at different materials and exploratory arrangements, we investigate the fundamental components, qualities, and expected uses of this charming peculiarity. The trial perceptions examined in this study shed light on the way materials behave when exposed to cyclic freezing, consequently adding to a more profound comprehension of the field of strong mechanics.

Presentation

Strong mechanics incorporates the investigation of how materials respond to outer powers and disfigurements. The term "periodic freezing," which refers to the recurrent formation and melting of ice crystals within a solid material, is a less well-known aspect of solid mechanics. A fascinating area for research is this cyclic freezing process, which results in distinct structural changes and mechanical properties. In this review, we present an outline of trial perceptions connected with occasional freezing and its suggestions.

Test Techniques

To research occasional freezing, different exploratory strategies have been utilized. The selection of materials, test planning methods, and freezing conditions assume essential roles in catching the peculiarity. Specialists have used controlled cooling techniques, for example, temperature-controlled chambers and fluid nitrogen showers, to actuate freezing in various materials. Techniques like thermal analysis, X-ray diffraction, and microscopy have been used to make observations. During freezing cycles, these methods enable the characterization of structural changes, crystal formation, and mechanical properties.

Materials Under Study A variety of materials have been examined to learn more about periodic freezing. Different types of metals, polymers, ceramics, and composites have responded to cyclic freezing. Every material has novel properties and microstructural attributes that impact its freezing conductivity. Scientists have zeroed in on understanding the impact of variables like creation, stage changes, porosity, and warm conductivity on the occasional freezing process.

Characteristics of periodic freezing Several key characteristics of periodic freezing have been discovered through experimental observations. First and foremost, the cyclic freezing process causes changes in the microstructure, such as the formation of ice crystals and their subsequent melting. These progressions can bring about adjustments in material properties like mechanical strength and thermal conductivity. Second, the material develops stress fields as a result of the freezing cycles, which can have an impact on its structural integrity. Thirdly, the material's overall behavior is significantly influenced by the frequency and length of the freezing cycles.

Mechanical Properties Periodic freezing can significantly alter a material's mechanical properties. The effects of cyclic freezing on stiffness, ductility, and fracture toughness have been demonstrated experimentally. The presence of ice crystals and their effect on the material's microstructure are to blame for these changes. Furthermore, the rehashed development and liquefying of ice precious stones present cyclic stacking on the material, which can prompt weariness and disappointment after some time.

Applications and Potential Benefits: Recognising the behavior of materials subjected to periodic freezing has significant ramifications for several different fields. The creation of materials that can withstand extreme temperature variations is one potential application. Also, occasional freezing might be tackled to prompt controlled microstructural changes in materials for custom-made mechanical properties. The discoveries from exploratory perceptions can likewise contribute to progress in fields like cryogenic design, biomaterials, and energy capacity frameworks.

Difficulties and Future Bearings

Although huge headway has been made in understanding occasional freezing, a few difficulties remain. Accomplishing a more profound comprehension of the fundamental instruments and precisely foreseeing the way materials behave under cyclic freezing conditions are regions that require further exploration. Moreover, creating normalized trial conventions and investigating new strategies for portrayal would upgrade our capacity to examine and take advantage of this peculiarity behavior and properties of materials subjected to cyclic freezing have been better-understood thanks to experimental observations of periodic freezing in solid mechanics. The investigation of this phenomenon presents exciting opportunities for the creation of novel applications and the design of brand-new materials. To fully comprehend the mechanisms involved and address the difficulties associated with periodic freezing, additional research is required. In the end, a thorough comprehension of this phenomenon will make it possible to realize its potential advantages in a variety of scientific and technological fields.

Theoretical Models

Theoretical models are designed to describe and examine periodic solid mechanics freezing These models include different viewpoints, like warm extension, stress age, strain accumulation, and harm development. By analyzing these models, we plan to acquire bits of knowledge about the basic components of intermittent freezing and its effect on the primary reaction of strong materials.

Models of Thermal Expansion

One of the main causes of periodic freezing effects is thermal expansion. Thermal strains are caused when the dimensions of a solid material change in response to changes in temperature. The behavior of thermal expansion during periodic freezing has been described by several theoretical models.

One normally utilized model is the direct extension model, which accepts that the material goes through isotropic development or compression relative to the adjustment of temperature. The relationship between temperature and strain is described by this model using the coefficient of thermal expansion. Even though it is straightforward, the linear expansion model has difficulty capturing intricate thermal expansion patterns, particularly in anisotropic materials.

To defeat this restriction, further developed models, for example, the warm strain slope model, have been created. These models consider the spatial variety of temperature and integrate higher-request terms to catch the non-direct and anisotropic impacts of warm development. They offer a more accurate depiction of the behavior of deformation during periodic freezing.

Models for Creating Stress

Intermittent freezing can instigate stresses inside a strong material, prompting mechanical disappointment or harm. The coefficient of thermal expansion, temperature gradients, and material properties all have an impact on the onset of stress during freezing. To predict the magnitude and distribution of stress in response to cyclic freezing, theoretical models have been proposed.

One generally utilized model is the direct flexible pressure examination, which accepts that the material acts flexibly under warm stacking. This model joins the warm development impacts with the versatility of the material to work out the subsequent pressure field. However, time-dependent effects like creep and viscoelastic behavior are left out of the equation.

Viscoelastic or creep models are used to include time-dependent phenomena. The effects of loading rate on stress generation during freezing and the time-dependent behavior of materials are taken into account in these models. They are especially helpful while concentrating on materials with huge time-subordinate reactions, like polymers or certain metals.

Strain Gathering Models

Occasional freezing can prompt strain accumulation inside strong materials, which can bring about total harm or disappointment over the long haul. The material's response to cyclic freezing has been predicted, and the progressive accumulation of strains has been described using theoretical models.

The strain-based fatigue model, which relates the total strain to the number of thermal cycles, is one model that is frequently utilized. In this model, the material is assumed to undergo cyclic loading and fatigue life cycle, which is one model that is frequently utilized. In this model, the material is assumed to undergo cyclic loading and a fatigue life curve. The model predicts the onset of failure due to strain accumulation and provides insight into the fatigue life of the material by tracking the accumulated strain.

Moreover, harm mechanics models are utilized to catch the ever-evolving harm advancement during intermittent freezing. These models take into account how the material's mechanical properties change as a result of the onset and growth of microcracks or defects. By incorporating harm collection with strain history, these models offer a complete comprehension of the material's way of behaving and remaining helpful.

Coupled Thermo-Mechanical Models

To capture the full intricacy of occasional freezing peculiarities, hypothetical models that couple warm and mechanical reactions have been created. The purpose of these models is to simulate the interaction between changes in temperature, thermal expansion, the production of stress, the accumulation of strain, and the progression of damage.

A popular numerical method for creating coupled thermo-mechanical models is finite element analysis (FEA). Under cyclic freezing conditions, complex geometries and material behaviors can be simulated with FEA. By integrating the temperature-subordinate material properties and fitting constitutive conditions, FEA gives an incredible opportunity to concentrate on the coupled impacts of freezing on the underlying reaction.

PLASTIC DEFORMATION AND PERIODIC FREEZING

Interplay between Plastic Deformation and Periodic Freezing

Plastic deformity and freezing are two particular cycles that happen in different materials under various circumstances. Plastic disfigurement alludes to the super-durable change in shape or size of a material brought about by an outer power, while occasional freezing includes the change of a substance from a fluid to a strong state at customary spans. Even though they might appear to be unrelated, there are times when they interact with one another and have an impact on one another. Due to its implications for a variety of fields, including material science, geology, and engineering, this interaction between plastic deformation and periodic freezing has received a lot of attention in scientific research. This chapter investigates the instruments, impacts, and utilizations of this captivating interaction.

Mechanisms

- **Deformation of Plastic:** Plastic deformity happens when a material goes under pressure past its flexible breaking point, bringing about long-lasting changes in its shape or size. Atoms or molecules are reorganized as a result of this process, which involves the movement of dislocations in the crystal lattice. Plastic twisting can happen through different components, including slip, twinning, and disengagement climb. The capacity of a material to undergo plastic deformation influences its mechanical properties, such as its yield strength, ductility, and toughness.
- **Frequent Freezing:** Intermittent freezing alludes to the cyclic change of a substance from a fluid to a solid state under unambiguous circumstances, like temperature and tension. As the liquid cools and the atoms or molecules arrange themselves into a structured lattice, this typically results in the formation of crystals. External factors, such as cyclic temperature variations, cause repeated cycles of solidification and melting, resulting in the periodic nature of freezing.

Collaborations and Impacts

- **Plastic twisting incited freezing:** Plastic deformation can prompt nearby temperature changes inside a material, which might prompt occasional freezing in specific cases. The deformation of a material creates heat because of the inner erosion between its constituent particles. This confined temperature climb can make nearby districts arrive at the edge of freezing over, bringing about occasional cementing. This peculiarity has been seen in metalworking processes like virus rolling, where the rehashed twisting and limited warming can prompt the arrangement of fine-grained microstructures.
- **Deformation of plastic caused by freezing:** On the other hand, intermittent freezing can impact the plastic disfigurement conductivity of a material. At the point when a substance freezes, it goes through a volumetric development, applying tension to its

environmental factors. The material's mechanical properties and deformation behavior can be altered and stressed by this expansion. For example, the freezing of water in rocks can apply huge tension, prompting break spread and deformation in land developments.

Applications

- **Science of Materials:** In materials science, it is essential to comprehend how plastic deformation and periodic freezing interact with one another. Researchers can influence the microstructural evolution and mechanical properties of materials by modifying freezing conditions, such as temperature cycling. The creation of advanced alloys, composite materials, and coatings with individualized properties are just a few examples of the practical applications of this knowledge.
- **Geology:** In geology, the Earth's surface is shaped by the interaction of plastic deformation and periodic freezing. The freezing and defrosting of water inside rocks contribute to processes like ice wedging, which can break and deteriorate rock development. This mechanism is especially important in cold areas and mountainous environments, where rocks are weathered and eroded by repeated cycles of freezing and thawing.
- **Engineering:** The interchange between plastic deformation and intermittent freezing has suggestions for designing applications as well. Engineers can design structures that can withstand cyclic loading and temperature variations with the assistance of an understanding of the interaction between freezing and deformation. This information is significant for creating a sturdy framework in locales characterized by outrageous temperature variances, like extensions, pipelines, and seaward stages.

Future Headings

Even though we now know a lot more about how plastic deformation and periodic freezing interact, there are still a lot of things we can learn. Our comprehension of this interaction would be enhanced by additional investigations into the underlying mechanisms, such as the function of dislocation movement and crystal growth kinetics. Investigating the effects of various variables like strain rate, freezing rate, and material composition would also shed light on the intricate interactions that occur between these processes.

A fascinating area of scientific study with far-reaching repercussions is the interaction between periodic freezing and plastic deformation.

The microstructure, mechanical properties, and behavior of materials can all be affected by the mechanisms by which these processes interact with one another. The design of advanced materials and structures with enhanced performance and durability is made possible by comprehending this interaction, which has significant applications in materials science, geology, and engineering. We can anticipate discovering new insights and unlocking additional innovation and application potential as research in this area continues.

Experimental Techniques and Observations

- **Tests of Tension:** Ductile testing is a broadly utilized exploratory procedure to concentrate on the behavior of materials under strain. In this test, an example of the material is exposed to a hub load, expanding bit by bit until disappointment happens. Important mechanical properties like ductility, ultimate tensile strength, and yield strength can be determined during the test. The elastic and plastic deformation, strain hardening, and fracture behavior of materials can all be learned from tensile testing.
- **Pressure Testing:** A material specimen is subjected to a compressive load during compression testing. The behavior of materials under compressive forces, which can cause plastic deformation, can be studied with this method. The study of materials' deformation mechanisms and failure modes under various loading conditions can particularly benefit from compression testing.

Testing for Bending and Flexure

Twisting and flexural testing are ordinarily utilized to assess the mechanical properties of materials, particularly their protection from bowing burdens. A material specimen is subjected to a bending moment during these tests, either in a three-point or four-point bending configuration. The subsequent deformity and stress circulation give bits of knowledge into the material's solidarity, firmness, and break conduct.

- **Microscopy Strategies:** The microstructural changes that occur during plastic deformation can only be observed and characterized using techniques like optical microscopy, scanning electron microscopy (SEM), and transmission electron microscopy (TEM). Optical microscopy takes into consideration visible perceptions of the disfigured material, while SEM and TEM give high-goal imaging and examination of the microstructure at the nanoscale.
- **X-beam diffraction (XRD):** XRD is a strong method for examining crystallographic construction and leftover burdens in

materials. It is possible to obtain useful information regarding the lattice parameters, crystal orientation, and presence of defects by subjecting a material to X-rays and measuring the diffracted patterns. The study of the effects of plastic deformation on the crystal structure and phase transformations of the material can greatly benefit from XRD.

Observations of Periodic Freezing During the Deformation of Plastic

The phenomenon of periodic freezing, also known as the Portevin-Le Chatelier effect, occurs during plastic deformation in some alloys. It manifests as a series of abrupt drops in stress that cause irregular flow behavior. When certain alloying elements interact with dislocations and affect their mobility, materials that exhibit dynamic strain aging typically experience periodic freezing. Regarding periodic freezing, several observations have been made, including:

- **Oscillations of stress:** During occasional freezing, stress motions are seen in the pressure-strain bend of the material. These oscillations manifest as sudden drops in stress, which are followed by periods of stress recovery before the subsequent drop. The material's motion and interaction with dislocations are the causes of the stress drops.
- **Dependence on Temperature:** Occasional freezing is frequently impacted by temperature. Temperature fluctuations can alter the frequency and magnitude of stress drops. Stress drops are typically more pronounced and frequent at higher temperatures. The kinetics of dislocation interactions and the activation energy of the dynamic strain aging mechanism is responsible for this temperature dependence.
- **Sensitivity to Strain Rate:** The event and qualities of intermittent freezing are likewise affected by the strain rate applied to the material during plastic deformation. Stress drops and intermittent flow behavior tends to be more pronounced at higher strain rates. The kinetics of dynamic strain aging processes and the time required for dislocation interactions are related to the strain rate sensitivity of periodic freezing.
- **Effects on the Microstructure:** Periodic freezing is significantly influenced by the material's microstructural characteristics, such as grain boundaries, dislocation density, and precipitates. Interactions with these microstructural features and dislocation pile-ups have the potential to cause stress drops and contribute to the intermittent flow behavior that is observed during plastic deformation.

Exploratory procedures give important bits of knowledge into the systems and conduct of plastic deformation in materials. To comprehend a material's deformation behavior and determine its mechanical properties, tensile, compression, bending, and flexural testing are frequently used. Microstructural changes and crystallographic effects can be characterized using techniques like microscopy and X-ray diffraction. Moreover, the peculiarity of occasional freezing during plastic distortion has been seen in specific materials, prompting discontinuous stream conduct as described by pressure drops. Temperature, strain rate, and microstructural characteristics all have an impact on the occurrence and characteristics of periodic freezing, highlighting the complexity of plastic deformation behavior. To advance the field of materials science and engineering and design materials with improved mechanical properties, it is essential to comprehend these experimental methods and observations.

CASE STUDIES

Strong mechanics is worried about the investigation of how materials respond to outside powers and deformations. The permanent deformation of a material beyond its elastic limit is known as plastic deformation, and it is an essential component of solid mechanics. Two examples of the ideas of plastic deformation and periodic solid freezing are presented in this case study.

1st Case Study: In this case study, a metal rod subjected to a constant load is subjected to plastic deformation. At first, the bar is undeformed. The rod deforms elastically until it reaches its yield point as the load is applied. Past the yield point, the material enters the plastic disfigurement district, where super-durable deformities happen. The yield strength and strain-hardening behavior of the material influence the amount of plastic deformation.

To dissect this situation, we use pressure strain bends to imagine the material's reaction. There is a linear elastic region at the beginning of the stress-strain curve, followed by a yield point and a plastic deformation region at the end. The amount of energy the material takes in during plastic deformation is shown by the area under the stress-strain curve. It is essential to take note that plastic twisting is irreversible, and the material won't get back to its unique shape in the wake of dumping.

We examine the variables impacting plastic distortion, for example, the material's yield strength, strain rate, and temperature. We also investigate necking, which occurs when a small area of the material experiences excessive plastic deformation, reducing its cross-sectional area.

2nd Case Study: Occasional Freezing of a Strong: For this situation study, we examine the impacts of occasional freezing on the mechanical behavior of a strong. Materials may experience fatigue failure as a result of cyclic loadings, such as cycles of freezing and thawing. For example, substantial designs exposed to freezing and defrosting cycles can encounter microcracks and decreased strength over the long run.

Cyclically, we examine the freezing and thawing processes. At the point when a solid encounters freezing, water particles grow, applying interior weights to the material. The internal stresses are released upon thawing, potentially leading to cracking and material degradation. These effects are made worse by the cycle of freezing and thawing, which causes damage over time.

To relieve the hindering impacts of freezing and defrosting, different procedures can be utilized, like utilizing materials with low water retention, appropriate seepage frameworks, and defensive coatings. Understanding the way materials behave under cyclic stacking conditions is urgent for planning strong designs in conditions where freezing and defrosting are pervasive.

Theoretical Approaches

Predicting the response of materials in various applications and maximizing their performance requires theoretical approaches to plastic deformation and periodic freezing. This chapter talks about a portion of the hypothetical methodologies utilized in the investigation of plastic twisting and occasional freezing in strong mechanics.

1. Plastic Deformation Elastic and plastic behavior Deformation occurs when a solid material is subjected to an external force. The material returns to its original shape when the stress is removed because the deformation is elastic at low-stress levels. However, the material undergoes plastic deformation, resulting in a permanent change in its shape or size, when the stress exceeds a certain threshold known as yield stress. The theoretical approaches to plastic deformation aim to predict the behavior of materials beyond their yield point and explain the underlying mechanisms.
2. Dislocation Theory One of the fundamental theoretical frameworks for comprehending plastic deformation is the dislocation theory, which was developed by G.I. Taylor and F.C. Frank. As per this hypothesis, plastic deformity happens because of the development of line deserts called separations inside the gem cross-section of material. Dislocations can be moved by individual atoms moving or by groups of atoms moving together. Experiments and computer simulations have confirmed the mathematical description of dislocation motion provided by the theory.

3. Plastic Flow and Strain Hardening The ability of a material to continuously deform under applied stress without fracturing is referred to as plastic flow. Plastic deformation is frequently associated with this concept. Crystal lattice plane reorganization and dislocation movement are two characteristics of plastic flow. As the plastic twisting advances, the material will in general become more earnestly and more impervious to additional distortion, a peculiarity known as strain solidifying. To explain how materials harden under strain, theoretical models like the power-law and Voce hardening models have been developed.
4. Periodic Freezing Thermal Stresses and Freezing Periodic freezing is when a solid material experience repeated cycles of freezing and thawing. Thermal stresses can be created when a material freezes due to the volume expansion caused by the transition from a liquid to a solid phase. If these thermal stresses exceed the material's strength, they can cause mechanical deformation and even fracture. The purpose of theoretical approaches to periodic freezing is to comprehend the connection between changes in temperature, stresses caused by freezing, and the material's subsequent mechanical response.
5. Phase Field Modeling Computational phase field modeling has been extensively utilized to simulate and comprehend the behavior of materials during phase transformations, such as freezing. The solid-liquid interface can be simulated by using phase field models because it is represented as a diffuse boundary with a finite thickness. Researchers can forecast the growth and formation of ice crystals during freezing as well as the associated thermal stresses and deformations by resolving the appropriate equations governing the evolution of the phase field.
6. Thermoelasticity is a branch of solid mechanics that studies how materials' thermal and mechanical properties interact with one another. A mathematical description of the interaction between changes in temperature and the stresses and strains that result from those changes is provided by the thermoelasticity theoretical framework. By taking into account the temperature-subordinate material properties and warm extension coefficients, thermoelastic models can foresee the mechanical reaction of a material exposed to freezing and defrosting cycles. Understanding the effects of periodic freezing on the durability and integrity of structures and materials is made easier by these models.

RESULTS

On a solid mechanics setup, a series of experiments were carried out to investigate plastic deformation and the effects of periodic freezing. This study's material specimen was carefully chosen to represent typical engineering materials.

The chosen material was used to make samples with the same dimensions and starting conditions. The samples were meticulously machined to remove any defects or irregularities on the surfaces that could have affected how they deformed.

Setup of the Experiment

The exploratory arrangement consisted of a stacking framework fit for applying controlled powers to the example. Using strain gauges and extensometers, the corresponding deformations were measured as the forces were applied incrementally.

The samples were subjected to alternating cycles of freezing and thawing in a controlled environment to achieve periodic freezing. Before thawing the sample, the freezing cycles involved lowering the temperature to a predetermined level and maintaining that level for a predetermined amount of time.

The outcomes acquired from the trials uncovered a few key discoveries concerning plastic misshapen Ing and the impacts of occasional freezing on the strong mechanical framework. The following findings were observed:

- **Plastic twisting way of behaving:** Under expanding pressure, the material displayed both versatility and plastic distortion. At first, the material distorted flexibly, getting back to its unique shape upon the expulsion of the heap. Plastic deformation, on the other hand, took place above a certain stress threshold, resulting in permanent changes to the sample's shape.
- **Impact of Pressure Size:** The degree of plastic deformation was heavily influenced by the magnitude of the applied stress. The plastic strain was greater at higher stress levels, indicating a direct link between stress and deformation.
- **Mechanisms for Plastic Deformation:** The deformed samples had microstructural changes and dislocations, according to a microscopic examination. Plastic deformity happened because of the development and collaboration of disengagements inside the material's precious stone construction.

Impacts of Occasional Freezing

The intermittent freezing and defrosting cycles discernibly affected the plastic's twisting way of behaving. The sample's rate of plastic deformation was found to be affected by periodic freezing. The plastic strain was lower in samples that were frozen regularly than in samples that were not frozen.

- **Discussion:** The study's findings shed light on the plastic deformation behavior of materials and the possibility of periodic freezing affecting the deformation process. The key findings are summarised in the following discussion points:
- **Mechanisms for Plastic Deformation:** The presence of disengagements and microstructural changes shows that plastic misshapen Ing happens through the development and connection of separations inside the precious stone grid. Materials science theories of plasticity are supported by this finding.
- **The degree of stress's influence:** Higher applied stresses are correlated with greater plastic deformation, as evidenced by the direct relationship between stress magnitude and deformation. This information is fundamental for planning materials fit for enduring high-stress conditions without going through unreasonable plastic twisting.
- **Consequences of Repeated Freezing:** The fact that periodic freezing reduced the strain on the plastic in samples suggests that freezing may impede the plastic deformation process. This phenomenon may be attributed to a change in microstructural features or a change in dislocation mobility, but more research is needed to determine the exact mechanism.
- **Application possibilities:** The discoveries of this study can provide suggestions for different fields, including material design, fabrication, and foundational layout. The development of materials that are more resistant to deformation in environments with fluctuating temperatures can benefit from gaining an understanding of the effect that periodic freezing has on plastic deformation.

This study examined the plastic distortion conductivity of a strong mechanical framework and investigated the impacts of intermittent freezing. The significance of stress magnitude and the presence of dislocations in plastic deformation were brought to light by the findings. The study also demonstrated that the rate of plastic deformation can be affected by periodic freezing. The knowledge gained from this study could be used to improve materials for applications that require resistance to plastic deformation at different temperatures.

CONCLUSION

In conclusion, understanding the behavior of materials under stress relies heavily on solid mechanics' plastic deformation and periodic freezing phenomena. Periodic freezing is the repeated freezing and thawing of a material under cyclic loading conditions, whereas plastic deformation is the irreversible change in a material's shape or size when it is subjected to external forces beyond its elastic limit. These cycles have huge ramifications for different design applications and have been broadly examined and dissected.

The movement of dislocations within a material's crystal lattice causes plastic deformation. At the point when stress is applied, disengagements begin to move, making the material go through plastic twisting. Depending on the forces applied and the properties of the material, this deformation can take the form of bending, stretching, or shearing. To ensure that structures and components can withstand the anticipated loads and maintain their integrity, plastic deformation must be taken into consideration.

The development of various mathematical models and theories, such as the von Mises yield criterion and strain hardening models, which assist in predicting the behavior of materials under various loading conditions, has been facilitated by our comprehension of plastic deformation mechanisms. Engineers can now design structures with a sufficient safety margin and make optimal use of materials thanks to these models, which have demonstrated their usefulness in engineering applications.

In contrast, periodic freezing involves the cyclic loading and unloading of material, resulting in repeated cycles of freezing and thawing. When materials are subjected to alternating loads or in locations with significant temperature variations, this phenomenon is especially relevant. The material's properties can be altered as a result of the cycle's repeated freezing and thawing, such as decreased strength, increased brittleness, and the onset of cracks or fractures. In applications like pavement design, where freeze-thaw cycles can cause significant damage, it is essential to take into consideration the effects of periodic freezing.

Engineers employ a variety of methods and strategies to mitigate the negative effects of plastic deformation and periodic freezing. Materials with high ductility and strength are frequently used for plastic deformation because they can resist deformation and withstand higher loads without failing. The material's resistance to plastic deformation can also be improved through the use of heat treatment procedures like annealing or tempering.

On account of occasional freezing, engineers go to great lengths to limit the openness of materials to freeze-defrost cycles. Insulation and protective coatings, for instance, are utilized in the construction of infrastructure in cold regions to minimize temperature swings and the effects of freezing and thawing. In a similar vein, materials with enhanced fatigue resistance are selected to withstand repeated stress cycles in applications where cyclic loading cannot be avoided, such as the aerospace or automotive industries.

In outline, plastic deformity and occasional freezing are essential ideas in strong mechanics that altogether impact the way materials behave and execute under different stacking conditions. Understanding these peculiarities permits architects to configure designs and parts that can endure outer powers and natural elements, guaranteeing the security and solidity of designed frameworks. Our comprehension of plastic deformation and periodic freezing continues to grow as a result of ongoing research and advancements in material science and engineering. As a result, improved design strategies and novel application-specific solutions emerge.

4

Twisted Polymer Actuators and Optimization Design of a Solid Mechanics

INTRODUCTION

A promising class of smart materials with remarkable shape-changing capabilities in response to external stimuli has emerged as twisted polymer actuators. The unique combination of mechanical flexibility, lightweight Ness, and adaptability of these stimuli-responsive polymer actuators makes them highly desirable for solid mechanics applications. Innovative engineering solutions in fields like robotics, biomedical devices, and aerospace engineering have been made possible by twisted polymer actuators' capacity to undergo controlled large-scale deformations. However, the design of these actuators must be optimized for improved performance and efficiency to fully utilize their potential. The principles and mechanisms of twisted polymer actuation are the focus of this study, along with their applications in solid mechanics and the optimization design methods used to make them as effective as possible. By completely analyzing the plan boundaries, material properties, creation procedures, and enhancement techniques, this chapter expects to give important bits of knowledge and rules for the effective execution of contorted polymer actuators in common-sense applications.

Background and Motivation

Strong mechanics, a part of designing mechanics, manages the way strong materials behave under different stacking conditions. For the creation of cutting-edge engineering systems, it is essential to have the capability of manipulating and controlling the mechanical response of materials. For this purpose, a lot of traditional actuators, like electric motors and hydraulic cylinders, have been used. In any case, their size, weight, and restricted versatility present difficulties in specific applications where minimization, lightweight planning, and perplexing movement abilities are desired. The creation of smart materials and actuators that are capable of exhibiting unique mechanical properties and responding to external stimuli has garnered a growing amount of attention in recent years. Turned polymer actuators have emerged as a promising class of shrewd materials with extraordinary expected strength in mechanics. Most of the time, these actuators are made of stimuli-responsive polymers that can change significantly in shape in response to a variety of external triggers like temperature, light, or electric fields.

The need to investigate and improve the design of twisted polymer actuators for their effective use in solid mechanics applications is the impetus for this chapter. Although the idea of twisted polymer actuation holds promise, it is necessary to improve its performance, effectiveness, and dependability. Researchers and engineers can make use of the unique capabilities of twisted polymer actuators in a variety of fields by comprehending the underlying principles, mechanisms, and limitations of these actuators and exploring optimization design strategies. The expected utilizations of curved polymer actuators in strong mechanics are tremendous, going from delicate mechanical technology and wearable gadgets to aviation designs and clinical instruments. It is possible to create highly effective and custom-made twisted polymer actuators that meet the particular requirements of each application by optimizing their design parameters, material selection, fabrication techniques, and performance evaluation methods. By providing actuation solutions that are small, light, and adaptable, these advancements have the potential to transform engineering systems. The growing interest in twisted polymer actuators as smart materials for solid mechanics applications serves as the foundation for this chapter. It is possible to overcome existing limitations and unlock their full potential in various engineering domains by comprehending their principles, exploring their applications, and optimizing their design.

Objectives

Examine the fundamentals and workings of twisted polymer actuators. The underlying principles governing the actuation behavior of twisted polymer actuators will be examined in this chapter. A foundation for further optimization and design considerations will be established by comprehending the mechanisms by which these actuators undergo shape changes in response to external stimuli. Investigate the solid mechanics' applications of twisted polymer actuators. The potential uses of twisted polymer actuators in solid mechanics will be discussed in detail in this chapter. It will demonstrate the adaptability and efficiency of these actuators in various engineering fields by focusing on successful case studies and examples.

Analyze the difficulties and design considerations of twisted polymer actuators: The key design parameters and considerations that affect twisted polymer actuator performance and efficiency will be identified in this chapter. It will likewise talk about the difficulties related to their plan, like material choice, creation strategies, and control instruments. Look into ways to optimize the performance of twisted polymer actuators: The optimization strategies and methods used to boost the performance of twisted polymer actuators will be the focus of this chapter. Multi-objective optimization, genetic algorithms, mathematical modeling, simulation, and sensitivity analysis will all be examined. By dissecting these methodologies, the chapter plans to give bits of knowledge on how to streamline the design of curved polymer actuators for explicit applications.

Provide guidelines for using twisted polymer actuators effectively in solid mechanics: In light of the discoveries from the examination and investigation, the chapter will introduce useful rules and suggestions for the planning, manufacture, and coordination of wound polymer actuators in strong mechanics applications. These rules will help specialists and scientists use these actuators and expand their presentation in assorted design frameworks. To improve our understanding of twisted polymer actuators and provide useful insights for optimizing their design in solid mechanics by achieving these goals. It aims to make it easier to come up with creative engineering solutions that take advantage of the special capabilities of twisted polymer actuators and make changes in a variety of industries possible.

TWISTED POLYMER ACTUATORS: PRINCIPLES AND MECHANISMS

Introduction to Twisted Polymer Actuation

The phenomenon of certain polymers exhibiting significant shape changes or mechanical deformations in response to external stimuli is referred to as twisted polymer actuation. Variations in temperature, exposure to light, or the application of electric fields are all examples of these stimuli. Due to their unique capacity for large-scale deformation, twisted polymer actuators have attracted a lot of attention and are promising candidates for a variety of applications in solid mechanics. The activation conduct of these materials depends on the inborn properties of boost-responsive polymers, otherwise called brilliant polymers or shrewd gels, which can go through reversible changes in their actual properties upon openness to explicit triggers.

Types of Stimuli-responsive Polymers

Each type of stimuli-responsive polymer used in twisted polymer actuators is tailored to a specific application and trigger. The following are examples of commonly used stimulus-responsive polymers:

- **Polymers that respond to heat:** These polymers show changes in their shape or mechanical properties because of temperature variations. Shape memory polymers (SMPs) are one example, and they can return to their original shape after being deformed by a change in temperature.
- **Polymers that respond to light:** When these polymers are exposed to particular wavelengths of light, they undergo reversible changes in their properties, such as shape, stiffness, or adhesion. Photo-responsive fluid glasslike polymers and photothermal actuators are instances of photo-responsive polymers used in contorted polymer actuators.
- **Polymers that respond to electricity:** Shape shifts or mechanical deformations are the results of these polymers reacting to electric fields. In twisted polymer actuator designs, this category includes electroactive polymers (EAPs) and dielectric elastomers.

Actuation Mechanisms

To produce shape-shifting behavior, twisted polymer actuators make use of a variety of actuation mechanisms. The specific kind of polymer and the stimuli used determine the actuation mechanism. Some normal incitation systems include:

- **Mechanism for swelling and shrinking:** This component depends on the reversible expanding and deswelling of the polymer in light of the applied upgrade. Mechanical deformation caused by the change in volume results in actuation.
- **Shape memory impact:** Shape memory is a property of some polymers that allows them to temporarily deform and then return to their original form when exposed to the triggering stimulus. Programmable actuation behavior is made possible by this shape-memory effect.
- **Electrostatic or electrochemical activation:** Through electrostatic forces or electrochemical reactions induced by an applied electric field, electrically responsive polymers can deform or change shape.

Performance characteristics and limitations

Applications in solid mechanics are drawn to twisted polymer actuators due to their impressive performance characteristics. These attributes incorporate huge activation strains, quick reaction times, low power prerequisites, and the capacity to work in assorted ecological circumstances.

However, there are limitations to twisted polymer actuators that must be taken into consideration. Mechanical weakness, fatigue after many actuation cycles, environmental sensitivity, and difficulties in precisely controlling actuation behavior are all examples of these limitations.

To get the most out of these actuators in solid-mechanics applications, it is essential to comprehend the fundamentals and mechanisms of twisted polymer actuation. Researchers and engineers can use twisted polymer actuators to their full potential in a variety of engineering systems by utilizing the distinctive properties of stimuli-responsive polymers and investigating novel actuation mechanisms.

APPLICATIONS OF TWISTED POLYMER ACTUATORS IN SOLID MECHANICS

Overview of solid mechanics

A fundamental branch of mechanics that deals with the behavior of solid materials under various loading conditions is known as solid mechanics, solid body mechanics, or mechanics of solids. It incorporates the investigation of how solids respond to outside powers, distortions, and stress dissemination. Understanding and predicting the mechanical behavior of materials, designing structures and machines, and ensuring the safety and dependability of engineering systems all require solid mechanics.

- **Exhaustion and Stress:** Stress alludes to the interior powers inside a strong material coming about because of outer loads or applied powers. It is a proportion of the power of the inner powers following up on a particular region. On the other hand, the deformation or change in shape that a material experiences when subjected to stress is referred to as strain. It addresses the overall change in length, point, or volume of a strong because of applied powers.
- **Flexibility and Versatility:** A material's ability to deform under stress and return to its original shape after the stress has been removed is known as its elasticity. It is portrayed by the straight connection between anxiety inside the flexible furthest reaches of the material. Versatility, then again, alludes to the super-durable

disfigurement or change in shape that happens in a material past its flexible breaking point. Plastic mishaps are typically irreversible.

- **Strength and Criteria for Failure:** Strength is the capacity of a material to resist misshapen Ing or disappointment under applied loads. It is ordinarily portrayed by boundaries like yield strength, extreme strength, and crack durability. The conditions under which a material will fail or suffer structural damage are predicted using failure criteria, which are frequently defined in terms of stress or strain limits.
- **Displacement and Deformation:** The change in size or shape of a solid material caused by external forces is known as deformation. It can appear as direct, precise, or volumetric changes. Uprooting, then again, addresses the adjustment of position or development of explicit focuses inside a strong because of applied powers or misshapen Ings.
- **FEA, or finite element analysis:** In solid mechanics, the numerical technique known as finite element analysis is extensively utilized for analyzing and predicting the behavior of intricate structures. It includes separating a design into more modestly limited components, applying fitting conditions and limit conditions to every component, and tackling the subsequent arrangement of conditions to get the general behavior of the construction.

Numerous engineering fields, such as civil engineering, mechanical engineering, aerospace engineering, materials science, and biomechanics, rely heavily on solid mechanics. Engineers can design structures, machines, and components that can withstand anticipated loads and forces while maintaining structural integrity, safety, and optimal performance thanks to their comprehension of the mechanical behavior of materials. Strong mechanics standards are utilized in the examination and design of scaffolds, structures, airplanes, vehicles, clinical inserts, and a great many mechanical frameworks.

Engineers can evaluate the distributions of stress and strain within materials, as well as the factors that influence material failure, by studying solid mechanics. They can also make educated decisions regarding the selection of materials, component design, and structural optimization. In addition, advanced computational modeling techniques like finite element analysis, which engineers can use to simulate and predict the behavior of complex systems before their physical realization, are based on solid mechanics. Understanding the mechanical behavior of solids, creating dependable structures, and ensuring the safety and performance of engineering systems all require solid mechanics. Engineers can make more effective and dependable engineering solutions by applying stress, strain, deformation, and failure principles to material selection, structural design, and optimization.

Potential Applications of Twisted Polymer Actuators

Twisted polymer actuators are suitable for a wide range of industries due to their unique shape-changing and mechanical deformation capabilities. Twisted polymer actuators could be used in the following scenarios:

- **Delicate advanced mechanics:** In soft robotic systems where flexible and adaptable actuation is desired, twisted polymer actuators can be utilized. These actuators may make it possible for robots to imitate natural movements and interactions, such as grasping, crawling, and dexterous gripping of objects. Twisted polymer actuators are ideal for soft robotic applications due to their lightweight and compact design.
- **Biomedical Gadgets:** Biomedical engineering may benefit from the use of twisted polymer actuators. They can be utilized in clinical gadgets and inserts, like fake muscles, drug conveyance frameworks, and assistive gadgets. These actuators can give controlled and exact developments, empowering the advancement of negligibly obtrusive, careful apparatuses and prosthetic gadgets that intently copy human movement.
- **Optics and Photonics:** The shape-changing properties of bent polymer actuators make them appropriate for applications in optics and photonics. When the polymer is actuated, it can be used in adaptive lenses, tuneable optics, and optical switches to change the focal length, aperture, or refractive index. Contorted polymer actuators can provide a dynamic and responsive control component in optical frameworks.
- **Aeronautics and space travel:** Turned polymer actuators can track down applications in the aviation and flying industries. When these actuators are actuated, they can be used to morph wings, flaps, and control surfaces to alter the aircraft's aerodynamic properties. Bent polymer actuators can further develop mobility, effectiveness, and haul in airplane frameworks.
- **Technology for Wearables:** The lightweight and adaptable nature of bent polymer actuators makes them reasonable for wearable innovation applications. They can be incorporated into wearable gadgets like smart materials, exoskeletons, and haptic input frameworks. In wearable devices, twisted polymer actuators can enhance user comfort, mobility, and interaction by providing personalized and adaptable functionalities.

- **Systems with a Lab-on-a-Chip and microfluidics:** In both lab-on-a-chip systems and microfluidic devices, twisted polymer actuators can be utilized. They can precisely manipulate and control microscale fluids thanks to their ability to control fluid flow, valve operations, and micropump functions. Biomedical diagnostics, chemical analysis, and microfluidic lab processes could all benefit from these actuators.
- **Architecture and Active Structures:** Turned polymer actuators can be used in dynamic designs and compositional plans. They can make architectural elements that change shape, adaptive facades, and deployable structures possible. The activation of bent polymer actuators considers dynamic changes in the construction, empowering energy-effective and responsive structural plans.
- **Energy Endlessly Reaping:** The misshapen Ings and mechanical movements of wound polymer actuators can be tackled for energy endlessly collecting applications. Self-powered sensors, wearable electronics, and energy-efficient systems could all benefit from these actuators' ability to convert mechanical vibrations, motions, or deformations into electrical energy.

Twisted polymer actuators have a wide range of potential uses, some of which are as follows: Utilizing these actuators' unique capabilities in a variety of industries, new and inventive applications are likely to emerge as research and development in this field continue to advance.

Case Studies: Examples of Successful Applications

- **Robotic Soft Gripper:** Scientists at Harvard College developed a delicate mechanical gripper utilizing contorted polymer actuators. The gripper was designed for delicate and precise grasping of items of varying shapes and sizes. The gripper was able to change its shape and conform to the objects thanks to the actuators, resulting in a secure yet gentle grip. Twisted polymer actuators' potential to achieve dexterity and versatility in soft robotics was demonstrated in this application.
- **Shape-Transforming Designs for Aviation:** Shape-shifting structures for aerospace applications were developed at the Langley Research Centre of NASA using twisted polymer actuators. Because the actuators were incorporated into the structures of the wings, it was possible to dynamically alter the shape and curvature of the wings during flight. This ability streamlined execution, improved mobility, and decreased drag. The utilization of bent polymer actuators in aviation exhibited their true capacity for transforming structures for advanced flight execution.

- **Dynamic Contact Focal Points:** Twisted polymer actuators were used in the creation of active contact lenses by researchers at the University of California, San Diego. The actuators, which were incorporated into the lenses, made it possible to dynamically alter the focal power. The lenses could change their curvature by applying a small electric current, allowing for variable focusing capabilities. This application displayed the capability of curved polymer actuators in versatile optics and ophthalmic gadgets.
- **Biomimetic Soft Devices:** Using twisted polymer actuators, researchers at Harvard University's Wyss Institute for Biologically Inspired Engineering created a soft biomimetic device. The device demonstrated swimming-like movements and imitated the movement of stingray fins. To replicate the intricate undulating motion, the actuators provided the necessary controllability and adaptability. This application featured the capability of wound polymer actuators in making bio-motivated delicate mechanical frameworks.
- **Transforming Engineering:** Twisted polymer actuators were investigated for their potential application in morphing architecture by researchers at the Swiss Federal Institute of Technology (Zurich). Using actuators to direct the movement of shading elements, they created a prototype of an adaptive shading system for buildings. The actuators empowered continuous changes in the concealing framework, improving energy productivity and solace inside the structure. Twisted polymer actuators' potential for energy-efficient and responsive architectural solutions was demonstrated in this application.

These case studies show how twisted polymer actuators have been successfully used in a variety of applications, including architecture, biomedical devices, robotics, aerospace, optics, and aerospace. They feature the flexibility, versatility, and extraordinary capacities of wound polymer actuators in accomplishing creative and practical arrangements.

OPTIMIZATION DESIGN OF TWISTED POLYMER ACTUATORS

Design Considerations and Challenges

To ensure their optimal performance and dependability, designing twisted polymer actuators requires addressing several considerations and difficulties. Twisted polymer actuator design considerations and challenges include the following:

- **Material Determination:** Picking the right responsive polymer is critical for accomplishing the ideal activation conduct. Factors such as the ideal reaction time, activation range, mechanical strength, and natural similarity should be considered. Various polymers have shifting properties and execution qualities, and choosing the most reasonable material for a particular application is basic.

- **A Mechanism for Actuation:** The desired actuation behavior and the intended application both influence the selection and design of the actuation mechanism. Grasping the hidden standards and systems of activation, for example, enlarging deswelling, shape memory impact, or electrostatic or electrochemical incitation, is significant for planning the actuator's construction and controlling its reaction.
- **Design of the Structure:** The performance of the twisted polymer actuator is significantly influenced by its geometric layout and design. Contemplations like the shape, size, and plan of the polymer components, as well as the underlying scaffoldings and limitations, should be painstakingly intended to accomplish the ideal activation conduct and mechanical solidness.
- **Connections for electricity:** On account of electrically responsive wound polymer actuators, the plan ought to incorporate arrangements for electrical associations with the vital electric fields. To guarantee efficient actuation and minimize energy losses, the electrical connections need to be dependable, effective, and compatible with the actuator material.
- **Systems for Control and Feedback:** For precise actuation and achieving the desired response, it is essential to develop efficient control and feedback systems. Planning appropriate control calculations, sensors, and criticism systems to screen and change the actuator's conduct progressively can upgrade its exhibition and empower progressed functionalities.
- **Fatigue and Durability:** Over numerous actuation cycles, twisted polymer actuators may wear out and fatigue, resulting in decreased performance or failure. It is essential, to guarantee the actuators' long-term reliability and durability, and to design them using materials and structural components that can withstand cyclic loading.
- **Compatibility and Integrity:** Twisted polymer actuators typically require integration into larger devices or systems. For the system's overall functionality and performance, it is essential to ensure compatibility and seamless integration with other components like power sources, control electronics, or mechanical structures.
- **Scalability and manufacturing:** It's important to think about making twisted polymer actuators that can be made quickly and

cheaply in large quantities. To make it possible for the actuators to be produced at a low cost and used in a wide range of applications, it is necessary to take into account the manufacturing methods' scalability, repeatability of the processes, and fabrication techniques.

- **Factors from the outside:** The presentation and solidity of curved polymer actuators can be affected by natural factors like temperature, dampness, and openness to synthetic substances. Planning the actuators to endure and work in unambiguous ecological circumstances is fundamental for their solid exhibition and life span.
- **Considerations for Safety:** Any actuator system's design must consider safety. To guarantee the safe operation of twisted polymer actuators in real-world applications, it is essential to take into consideration aspects like mechanical integrity, electrical insulation, and risk mitigation in the event of failure or malfunction.

Interdisciplinary approaches involving expertise in materials science, mechanics, control systems, and manufacturing processes are required to address these design considerations and challenges. Through careful planning and streamlining, the maximum capacity of curved polymer actuators can be achieved, empowering their effective combination into different design frameworks.

Design Parameters and their Influence

Planning wound polymer actuators includes characterizing different boundaries that impact their exhibition, incitation conduct, and general usefulness. Here are some key plan boundaries and their effects on curved polymer actuators:

- **Polymer Choice:** Because it has a direct impact on the actuation behavior, response time, mechanical properties, and environmental compatibility of the actuator, selecting the appropriate polymer material is crucial. Stimuli-responsive properties like thermal, chemical, or electrical responsiveness are unique to each polymer. The choice of the polymer will determine the incitation system and the general presentation of the actuator.
- **Cross-Sectional Math:** The twisted polymer actuator's actuation behavior is significantly influenced by its geometric design as well as its cross-sectional shape. Factors like the width, length, and plan of the polymer components impact the actuator's reaction to outside upgrades, the activation range, and the power produced. The actuator's mechanical stability and adaptability are also affected by its geometry.

- **Pre-twist or pre-strain:** A crucial design parameter is whether the polymer actuator should be subjected to pre-strain or pre-twisting before actuation. Pre-strain or pre-bend presents flexible energy in the actuator, which is delivered during activation, empowering shape change or movement. The extent of the pre-strain or pre-contort determines the actuator's underlying state and the scope of twisting that is reachable.
- **The Actuation Band:** The activation range alludes to the degree of disfigurement or shape change that the wound polymer actuator can accomplish. It is influenced by the applied stimulus, the cross-sectional geometry, and the polymer material. To meet the needs of the intended application, it is essential to design for a particular actuation range.
- **Torque or Actuation Force:** The power or force created by the curved polymer actuator is another essential plan boundary. It determines whether the actuator can exert mechanical force or carry out a particular function. The actuation stimulus, as well as polymer parameters like its length, cross-sectional area, and mechanical properties, can have an impact on the force or torque.
- **Reaction Time:** The reaction season of the bent polymer actuator alludes to the time taken to accomplish the ideal incitation or shape change upon the use of the upgrade. It is influenced by the stimulus intensity, the actuator's geometry, and the polymer's viscoelastic properties. For applications that necessitate quick or precise actuation, it is essential to plan for a specific response time when designing.
- **Durability and mechanical stability:** Twisted polymer actuators can only function effectively if their mechanical stability and durability are maintained. The actuator's capacity to withstand repeated actuation cycles without fatigue or failure is influenced by parameters such as the polymer material, cross-sectional design, and structural support. The actuation amplitude, cyclic loading, and fatigue properties of the material all need to be taken into account when designing for mechanical stability.
- **Consumption of Power:** Twisted polymer actuators' power consumption is an important design factor, especially in applications that require energy-efficient requirements or a limited power supply. Power consumption is influenced by actuator parameters like actuation force, response time, and energy losses during actuation. Planning productive actuator designs and control frameworks can assist with limiting power necessities.
- **Compatibility and Integrity:** Plan boundaries ought to be considered to guarantee similarity and a consistent mix of the

curved polymer actuator with different parts or frameworks. Factors like electrical associations, mechanical points of interaction, and similarity with control hardware or power sources should be addressed to empower the actuator's effective combination within the general framework.

- **Environmental Factors to Consider:** For twisted polymer actuators to function reliably, they must be designed to withstand specific environmental conditions.

Material Selection and Characterization

Material choice is a basic part of planning bent polymer actuators, as it directly impacts their incitation conduct, mechanical properties, and execution. For twisted polymer actuator material selection and characterization, the following are important considerations:

- **Properties that Respond to Stimuli:** The actuation of twisted polymer actuators is dependent on the polymer's response to stimuli. Different polymers respond differently to temperature, pH, light, and electrical fields from the outside. Grasping the ideal incitation system and choosing a polymer with the appropriate improvements in responsive properties is essential.
- **Physical Characteristics:** The actuator's performance and durability are significantly influenced by the polymer's mechanical properties. The actuator's capacity to deform, recover, withstand multiple actuation cycles, and support mechanical loads is determined by parameters like stiffness, elasticity, and tensile strength. The mechanical properties are evaluated using material characterization methods like dynamic mechanical analysis, tensile testing, and compression testing.
- **Enlarging Conduct:** Swelling and deswelling of the polymer material are important actuation factors in some twisted polymer actuators. The polymer's molecular structure and interactions with the environment influence swelling behavior. For actuator design, it is essential to comprehend and characterize the swelling behavior, including the degree of swelling, response time, and reversibility.
- **Compatibility with the environment:** For long-term actuator performance, the polymer material must be compatible with the operating environment. It is necessary to take into consideration things like resistance to temperature, chemical resistance, and humidity tolerance. The actuator's reliability and stability in real-world applications are helped by material testing under relevant environmental conditions.

- **Biocompatibility:** Biocompatibility is essential in biomedical applications where twisted polymer actuators may come into contact with living tissues or fluids. The material ought to be non-poisonous, non-aggravating, and viable with organic frameworks. The material's suitability for biomedical applications is evaluated using biocompatibility tests like cytotoxicity assays and in vitro tissue culture studies.
- **Handling and Manufacturing:** Important considerations include the material's processability and compatibility with fabrication methods. Depending on the actuator structure that is desired, the material should be able to be processed using extrusion, casting, or 3D printing. The actuators' reproducibility and efficient production are ensured by knowing the material's processability.

Fabrication Techniques and Processes

Twisted polymer actuators' material properties and actuation behavior are evaluated using a variety of characterization methods. These are some:

- **Thermal Examination:** The thermal properties of the polymer material, such as the glass transition temperature or thermal stability, are examined using methods like differential scanning calorimetry (DSC) and thermogravimetric analysis (TGA).
- **Spectroscopic Investigation:** Strategies like Fourier-change infrared spectroscopy (FTIR) or Raman spectroscopy assist with breaking down the compound construction and creation of the polymer material.
- **Mechanical Testing:** The material's mechanical properties, such as Young's modulus, tensile strength, and strain behavior, are evaluated through tensile testing, compression testing, or dynamic mechanical analysis (DMA).
- **The characteristics of swelling:** gravimetric analysis, optical microscopy, or spectroscopic methods are used to measure the degree of swelling and response kinetics to evaluate swelling behavior.
- **Electrical Properties:** Electrical characterization methods, such as impedance spectroscopy or electrical conductivity measurements, can be utilized to evaluate the material's electrical properties if the actuator relies on an electrical stimulus. Material choice and portrayal include a mix of exploratory testing and hypothetical comprehension of polymer conduct.

Performance Evaluation and Testing

The creation of twisted polymer actuators to guarantee their usability, dependability, and suitability for particular applications. Here are a few critical parts of execution assessment and testing for turned polymer actuators:

- **Activation Portrayal:** It is necessary to conduct thorough characterizations of the twisted polymer actuators' actuation behavior. This includes estimating boundaries, for example, incitation range, reaction time, power or force produced, and energy utilization. Activation portrayal can be directed utilizing exploratory arrangements that apply the significant boost (e.g., temperature, pH, electrical field) and measure the subsequent incitation reaction.
- **Tests on the Mechanical:** Twisted polymer actuators' mechanical properties, durability, and structural integrity are evaluated through mechanical testing. Pliable testing, pressure testing, and cyclic testing are usually used to assess boundaries like rigidity, firmness, stretching at the break, and exhaustion obstruction. These tests guarantee that the actuators can endure mechanical loads and repeated incitation cycles without disappointment.
- **Testing in the Environment:** In their intended applications, twisted polymer actuators may be exposed to a variety of environmental conditions. Natural testing includes exposing the actuators to temperature variations, stickiness, compound substances, or other pertinent ecological variables to assess their presentation and solidness. This testing determines the actuator's appropriateness for explicit working circumstances and recognizes any likely impediments or debasement components.
- **Reliability and stability Over the Long Run:** Evaluating the drawn-out security and dependability of bent polymer actuators is critical for their common-sense use. To evaluate the actuator's performance over a longer period, accelerated aging tests, such as prolonged actuation cycles or exposure to harsher environmental conditions, can be carried out. The actuator's durability and dependability can be determined by monitoring parameters like its mechanical properties, response time, and actuation range.
- **Power consumption and efficiency:** It is essential to evaluate the power efficiency and energy consumption of twisted polymer actuators, particularly in situations where there is a limited power supply or energy restrictions. Testing the actuators under various working circumstances and estimating their power utilization can give bits of knowledge about their energy effectiveness. The actuator design and control strategies can be improved using this information to reduce power consumption.
- **Evaluation of biocompatibility (if necessary):** Biocompatibility testing is required if the twisted polymer actuators are intended

for biomedical applications. In vivo experiments, cytotoxicity testing, tissue compatibility studies, and other methods are used to evaluate the actuators' interactions with biological systems. When the actuators come into contact with living things, biocompatibility testing ensures that they won't harm anything or react with tissue.

- **Functional and system integration testing:** Twisted polymer actuators are frequently incorporated into more substantial devices or systems. Practical testing includes coordinating the actuators with different parts or subsystems and assessing the general framework's exhibition and usefulness. This testing guarantees that the actuators can successfully play out their planned errands inside the total framework.
- **Checking against the requirements of the design:** The performance of twisted polymer actuators is compared to the predetermined design requirements and specifications during the testing process. The actuators will be validated to ensure that they meet the application-specific functional and performance requirements.

Researchers and designers can identify and address any limitations, optimize the actuator design, and guarantee the reliability and effectiveness of twisted polymer actuators in real-world applications by carrying out extensive performance evaluation and testing.

OPTIMIZATION METHODS AND TECHNIQUES

Mathematical Modeling and Simulation

These apparatuses empower architects and scientists to figure out the basic standards, foresee the actuator's way of behaving, and investigate different plan setups before trial prototyping. Here is an outline of numerical demonstration and recreation concerning bent polymer actuators:

- **Behavior Modelling for Actuation:** Numerical models are created to depict the activation conductivity of curved polymer actuators. These models capture the connection between the boost (e.g., temperature, pH, electrical field) and the subsequent twisting or movement of the actuator. The actuation mechanism can be accurately represented using a variety of modeling approaches, such as constitutive equations, thermodynamic models, or multi-field coupling models.
- **Using a Constitutive Model:** The actuator's polymer's material behavior is described by constitutive models. These models capture the connection between stress, strain, and the improvements following up on the material. The decision to fit a constitutive model

depends upon particular polymer properties, like viscoelasticity, a non-direct way of behaving, or expanding qualities. The constitutive model's material parameters can be derived from the literature or experimental testing.

- **Finite Element Analysis (FEA):** Limited component examination is broadly used to recreate the mechanical behavior of turned polymer actuators. The actuator's geometry is broken down into smaller, finite elements using FEA, and the mechanical response equations are solved numerically. FEA empowers the expectation of misshapen Ings, stresses, and strains in complex actuator calculations and considers improvement studies to distinguish ideal plan setups.
- **Simulation of Multiphysics:** Multiple physical phenomena, such as thermal, electrical, or chemical fields, are frequently coupled with mechanical deformation in twisted polymer actuators. Multiphysics re-enactment consolidates the overseeing conditions of various fields into a bound-together model and settles them all the while. This makes it possible to investigate coupled effects like how the actuator's deformation is affected by temperature or electrical fields.
- **Control Framework Recreation:** The actuator's behavior can be modeled using simulation tools for the control system. Simulations can predict the actuator's response to various control inputs by combining the mathematical model of the actuator with control algorithms. The actuator's performance under various operating conditions can be predicted, and the control strategy can be improved as a result of this.
- **Enhancement of Parameters:** Parameter optimization is made possible by mathematical modeling and simulation to improve the actuator's performance. Optimization algorithms can be used to find the best actuator configurations that meet specific performance criteria by defining design variables and objective functions. As a result, actuator geometries that maximize actuation range, response time, or force generation can be designed and discovered.
- **Awareness Examination:** An awareness examination includes assessing what changes in input boundaries mean for the actuator's result. It helps to identify the most important parameters, comprehend how they affect the behavior of the actuator, and direct the design process. The robustness and sensitivity of the actuator's performance to changes in material properties or operating conditions can be evaluated through sensitivity analysis using mathematical models or simulation tools.

Engineers can reduce the need for costly and time-consuming experimental iterations by using mathematical modeling and simulation to explore a wide range of design options, optimize actuator performance, and explore a variety of design possibilities. They provide insights into the actuator's behavior, direct the design process, and accelerate the creation of twisted polymer actuators that are effective and dependable.

Multi-objective Optimization Approaches

When designing twisted polymer actuators, multiple competing goals must be taken into account, and multi-objective optimization methods are used. Rather than looking for a solitary ideal arrangement, multi-objective streamlining intends to find a bunch of arrangements that address the compromises between various targets. In the context of twisted polymer actuators, the following are some of the most frequently applied multi-objective optimization strategies:

- **Methods based on Pareto:** The goal of Pareto-based optimization techniques, such as evolutionary algorithms or techniques based on the Pareto front, is to identify a set of solutions that make up the Pareto front. The Pareto front is the ideal compromise between competing goals where achieving one would necessitate giving up another. These techniques create a different arrangement of arrangements that cover a scope of compromise choices, permitting chiefs to choose the most reasonable arrangement because of their inclinations.
- **Method of Weighted Sum:** By assigning weights to each objective, the weighted sum method makes multiple objectives into a single objective. The optimal solution is then minimized or maximized by the weighted sum of the objectives. Weights that reflect the relative importance of each goal are needed for this method. Designers can explore a variety of objectives' trade-offs by adjusting the weights.
- **Techniques for Handling Constraints:** The design of twisted polymer actuators may sometimes be constrained by performance requirements, geometric constraints, or restrictions on the properties of the material. Imperative taking care methods, for example, limitation collection or punishment capabilities, can be utilized to integrate requirements into the multi-objective improvement process. These methods guarantee that the generated solutions optimize the objectives while meeting the specified constraints.
- **Hereditary Calculations:** Hereditary calculations (HC) are populace-based improvement calculations motivated by the course of regular determination. GAs evolves a population of solutions over generations by employing selection, crossover, and mutation

mechanisms. Numerous goals can be at the same time improved utilizing GAs by assessing the wellness of every arrangement in light of the targets. Exploration of the solution space and convergence toward the Pareto front are made possible by HCs.

- **Optimization of Particle Swarms:** Particle Swarm Optimization (PSO) is an optimization algorithm that is based on swarm intelligence and tries to replicate the actions of a flock of birds or a swarm of particles. A group of particles moves through the search space in PSO to find the best solutions. Every molecule's position addresses an answer, and its development is impacted by its own most popular position and the most popular places of adjoining particles. PSO has been applied to multi-objective improvement issues, including the plan of contorted polymer actuators.
- **Assisted Surrogate Optimization:** Based on a small number of evaluations, it is possible to approximate the performance of twisted polymer actuators using substitute models like polynomial regression, neural networks, or Gaussian processes. To explore the design space and locate promising solutions, surrogate-assisted optimization combines the effectiveness of surrogate models with multi-objective optimization algorithms. This method makes it possible to explore the design space to a greater extent and reduces the amount of computation required to evaluate intricate models.

When designing twisted polymer actuators, these multi-objective optimization methods offer designers a set of trade-off solutions that strike a balance between competing goals. By taking into account different goals at the same time, these methodologies work with the investigation of plan options and assist fashioners with settling on informed choices given their inclinations and plan necessities.

Genetic Algorithms and Evolutionary Techniques

They are broadly utilized in different fields, including the design and enhancement of wound polymer actuators. An overview of genetic algorithms and evolutionary methods can be found here:

- **Basics of Genetic Algorithms:** The principles of natural selection and evolution underpin the operation of genetic algorithms. The calculation begins with a populace of possible arrangements, addressed as people or chromosomes. Every individual encodes an answer by utilizing a bunch of boundaries or factors. The calculation then iteratively applies choice, hybrid, and change administrators to create new ages of people.
- **Wellness Assessment:** In hereditary calculations, the wellness of every individual is assessed because of a wellness capability

that addresses the goal to be streamlined. The wellness capability estimates the quality or execution of a single arrangement. The higher the fitness of the solution, the greater its likelihood of being chosen for reproduction and transferring its genetic material to subsequent generations.

- **Selection:** The process of selecting individuals from the current population based on their fitness values for reproduction is known as selection. High-wellness people have a higher likelihood of being chosen, while low-wellness people might be disposed of or have a lower opportunity of being picked. The selection process can be guided by employing a variety of selection methods, including rank-based selection, tournament selection, and roulette wheel selection.
- **Crossover:** The process of creating offspring by combining the genetic information of two parents is known as crossover. This is comparable to the natural process of reproduction. In hereditary calculations, hybrid commonly includes trading hereditary material (boundaries or factors) between parent people at explicit hybrid places. This makes new arrangements that acquire attributes from the two guardians.
- **Mutation:** Transformation acquaints arbitrary changes or varieties with individual arrangements. It assists with bringing variety to the populace and investigating various districts of the arrangement space. In hereditary calculations, change arbitrarily alters the hereditary material (boundaries or factors) of chosen people, prompting possibly better-than-ever arrangements.
- **Developmental Administrators:** By iteratively applying determination, hybrid, and transformation activities, hereditary calculations mimic the course of normal advancement over different ages. Thanks to this iterative procedure, the algorithm can search the solution space for solutions that are either optimal or very close to the optimal course of normal advancement over different ages. Thanks to this iterative procedure, the algorithm can search the solution space for solutions that are either optimal or very close to optimal. The calculation goes on until an end basis is met, like the greatest number of ages or combinations of the populace.
- **Enhancement of Parameters:** Parameter optimization problems in which the goal is to find the best values for a set of parameters that define a system or model are ideal candidates for genetic algorithms. On account of curved polymer actuators, hereditary calculations can be utilized to find the ideal mix of plan boundaries that amplify the actuator's exhibition or meet explicit plan targets.

- **Robustness and Discovery:** Hereditary calculations display heartiness and investigation abilities. The term "robustness" refers to a person's capacity to deal with noisy or imperfect fitness assessments and seek out effective solutions in such circumstances. The ability of the algorithm to explore various areas of the solution space, even those that were not initially explored, to potentially discover better solutions is referred to as exploration.

Twisted polymer actuator optimization has been achieved with success using genetic algorithms and evolutionary methods. They give you a flexible and effective way to look at different design options, think about multiple goals, and find the best or close to the best solutions. By copying the course of regular development, these strategies give a strong streamlining structure to the plan and improvement of wound polymer actuators.

Sensitivity Analysis and Parametric Studies

They assist with grasping the impact of information boundaries on the exhibition of the actuators and distinguishing basic boundaries that essentially influence the actuator's way of behaving. Here is an outline of responsiveness examinations and parametric investigations:

- **Awareness Examination:** The process of quantifying the relationship between changes in the output or performance of the twisted polymer actuators and variations in the input parameters is known as sensitivity analysis. It provides insight into the system's sensitivity to parameter variations and helps identify which parameters have the greatest impact on the actuator's behavior. Parameter sensitivity can be evaluated using sensitivity analysis methods like one-factor-at-a-time (OFAT), factorial design, or variance-based methods like Sobol's indices.
- **Variation in Parameter:** In parametric examinations, various qualities or reaches are assigned to the information boundaries of the turned polymer actuators, and their consequences for the actuator's exhibition are assessed. By efficiently shifting one or various boundaries while keeping others steady, the review gives an exhaustive comprehension of how the actuator answers changes in various plan factors. Parametric investigations can be performed utilizing mathematical reproductions, trial testing, or a blend of both.
- **Identifying Important Parameters:** Awareness examination and parametric investigations assist with recognizing the basic boundaries that altogether impact the actuator's exhibition. During the design and optimization process, engineers and researchers can prioritize their focus on the most influential factors by quantifying

the sensitivity or ranking the importance of various parameters. The selection of key parameters to be controlled or optimized to achieve desired actuator performance is guided by this data.

- **Streamlining and Plan Improvement:** Responsiveness examination and parametric investigations give significant data to streamlining and plan improvement. Optimization algorithms, such as gradient-based methods or genetic algorithms, can be used to determine the ideal values or ranges for the critical parameters once they have been identified. The actuator's performance can be improved and specific design objectives more effectively achieved by optimizing the influential parameters.
- **Analysis of Robustness:** Awareness investigations and parametric examinations can likewise survey the robustness of the contorted polymer actuators. Engineers can identify parameter ranges that result in robust operation and acceptable performance levels by evaluating the effects of parameter variations on the actuator's performance. This makes it easier to create actuators that can withstand changes in operating conditions or material properties, ensuring that they will perform consistently and reliably in real-world applications.
- **Guidelines and insights for design:** Design guidelines and insights are provided by sensitivity analysis and parametric studies. They assist with figuring out the connections between input boundaries, yield reactions, and the fundamental physical science of bent polymer actuators. These experiences can educate the advancement regarding configuration rules, rules, and best practices for upgrading actuator execution, directing plan cycles, and speeding up the plan interaction.

Responsiveness examination and parametric investigations are useful assets for figuring out the way curved polymer actuators behave, distinguishing basic boundaries, and working on their plan and execution. Engineers can improve the reliability and efficacy of twisted polymer actuators in a variety of applications, optimize designs, and make well-informed decisions by systematically varying parameters and quantifying their impact.

Case Studies: Optimization Design of Twisted Polymer Actuators

The ideal blend of plan boundaries and factors to boost their presentation, proficiency, and usefulness. The objective is to meet specific design goals while taking into account constraints and trade-offs. Here are key stages and contemplations in the enhancement plan of bent polymer actuators:

- **Define the goals of the design:** Obviously, characterize the plan goals because of the particular application necessities. Maximizing actuation range, speeding up response times, maximizing force output, increasing energy efficiency, or meeting specific performance goals are all examples of design objectives.
- **Determine the design parameters:** Identify the actuator's performance-influencing design parameters and variables. These may incorporate material properties (e.g., polymer type, firmness, viscoelasticity), mathematical boundaries (e.g., aspects, thickness, shape), incitation conditions (e.g., temperature, electrical field), or control boundaries (e.g., input voltage, timing).
- **Set design restrictions:** identify any constraints that should be taken into consideration when optimizing. Material restrictions, manufacturing constraints, geometric restrictions, or performance requirements are all examples of constraints. Constraints assist in ensuring that the optimized design is practical and feasible.
- **Create a logical model:** Create a mathematical model that accurately reflects the twisted polymer actuator's behavior. This model ought to capture the connection between the plan boundaries, the incitation instrument, and the ideal exhibition results. The model can be based on theories from thermodynamics, solid mechanics, or other relevant fields.
- **Choose a Method of Optimization:** Select an appropriate optimization technique in light of the design problem's complexity. Genetic algorithms, particle swarm optimization, gradient-based techniques, and surrogate-assisted optimization are all common approaches. Computational efficiency, convergence properties, and the capacity to deal with constraints are all important considerations.
- **Create the Function of the Objective:** Create an objective function that defines the optimization problem and quantifies the design objectives. The goal capability ought to be founded on the numerical model and mirror the ideal compromises between various targets. For multi-objective optimization, weighted sum methods or Pareto-based approaches may be used to strike a balance between competing goals.
- **Perform iterations of optimization:** Utilizing various combinations of design parameters, iteratively evaluate the objective function to carry out the optimization process. The optimization algorithm searches the design space for solutions that are either optimal or close to it. The objective function and constraints are used to evaluate a set of design variables that are generated with each iteration.

- **Assess and confirm solutions:** Assess the enhanced arrangements utilizing recreations, mathematical investigation, or exploratory testing. Based on the design objectives, constraints, and relevant performance metrics, evaluate the actuators' performance. To measure improvement, validate the optimized designs against desired specifications and compare them to initial or benchmark designs.
- **Evaluation of Robustness and Sensitivity:** To assess the robustness of the optimized design and comprehend the impact of parameter variations, conduct sensitivity analyses and robustness assessments. Determine the critical parameters that have a significant impact on the actuator's performance and guarantee that the optimized design is resistant to changes in material properties or operating conditions.
- **Improvement by Iteration:** Emphasise the improvement interaction in light of the assessment results, awareness examination, and plan experiences. Alter the objective function, modify the design parameters, or investigate alternative optimization strategies to refine the design. The optimization iterations should be repeated until the desired performance goals or design requirements are met.

Mathematical modeling, optimization algorithms, performance evaluation, and refinement are all components of the iterative, multidisciplinary optimization design process for twisted polymer actuators. The optimization process results in improved actuator designs that meet specific application requirements and maximize their performance potential by systematically exploring the design space and taking into consideration various design parameters and objectives.

Design Optimization for Enhanced Actuation Performance

Maximizing twisted polymer actuators' actuation range, response time, force output, energy efficiency, and other relevant performance metrics are all part of design optimization for improved actuation performance. For improved actuation performance, the design should be optimized in the following key ways:

- **Define the goals for the performance:** Based on the particular requirements of the application, clearly define the performance goals. These goals might incorporate boosting incitation relocation, speeding up, enhancing force age, limiting energy utilization, further developing dependability, or accomplishing explicit execution targets.
- **Determine the Design Parameters:** Determine the variables and design parameters that have a significant impact on the

actuation performance. Material properties (such as the type of polymer, stiffness, and thermal properties), geometric factors (such as dimensions, shape, and aspect ratio), electrode configuration, the actuation mechanism (such as the electric field and temperature), or control parameters are examples of these parameters.

- **Create a logical model:** Create a mathematical model that relates the design parameters to the desired performance outcomes and describes how the twisted polymer actuator behaves. Material properties, mechanical response, electro-active behavior, and any other relevant factors affecting actuation performance should be taken into account by the model. The model can be founded on strong mechanics, electro-mechanics, or other applicable speculations.
- **Create the Function of the Objective:** Make a goal capability that evaluates the ideal incitation execution and characterizes the advancement issue. The earlier-described performance objectives ought to be reflected in the objective function. For instance, it very well may be a blend of boosting incitation dislodging, limiting reaction time, and streamlining energy effectiveness. The goal capability fills in as the measurement to be enhanced during the planning cycle.
- **Choose a Method of Optimization:** Based on the complexity of the design problem and the resources available, select an appropriate optimization technique. Particle swarm optimization, gradient-based methods, genetic algorithms, and other methods of optimization are all options. Computational efficiency, the capacity to deal with constraints, and suitability for the problem at hand are all important considerations.
- **Perform iterations of optimization:** Utilizing various combinations of design parameters, iteratively evaluate the objective function to carry out the optimization process. The optimization algorithm searches the design space for solutions that are either optimal or close to it. Every emphasis creates a bunch of plan factors that are assessed because of the goal capability and any pertinent requirements.
- **Assess and confirm solutions:** Assess the enhanced arrangements utilizing recreations, mathematical investigation, or exploratory testing. Utilizing the established goals and pertinent performance metrics, evaluate the optimized designs' actuation performance. To measure improvement, validate the optimized designs against desired specifications and compare them to initial or benchmark designs.

- **Evaluation of Robustness and Sensitivity:** To comprehend the impact of parameter variations and ensure that the optimized design is resilient to variations in operating conditions or material properties, conduct sensitivity analyses and robustness assessments. Consider the sensitivity of critical parameters that have a significant impact on actuation performance during the design process.
- **Improvement by Iteration:** Emphasise the improvement interaction in light of the assessment results, awareness examination, and plan experiences. Adjust the design parameters, alter the objective function, investigate alternative optimization strategies, or take into account additional constraints to refine the design. The optimization iterations should be repeated until the desired actuation performance targets or design requirements are satisfied.
- **Approval and Execution Testing:** Approve the enhanced plan through thorough testing and execution assessment. Analyze the actuator's behavior experimentally and contrast the findings with the optimization process's predicted performance. This step guarantees that the advanced plan converts into further developed activation execution in common-sense applications. The optimization design process has the potential to improve the actuation performance of twisted polymer actuators if these steps are followed and the relevant design parameters and performance goals are taken into consideration. It allows designers to consider constraints and trade-offs while locating optimal or near-optimal solutions that maximize their performance capabilities.

Design Optimization for Improved Efficiency and Reliability

The twisted polymer actuators' energy efficiency must be maximized while ensuring consistent and reliable performance. The following are important things to think about and do to make the design more effective and reliable:

- **Define objectives for reliability and efficiency:** Obviously, characterize the effectiveness and dependability targets because of the particular application prerequisites. Effectiveness goals might incorporate boosting energy transformation proficiency, limiting power utilization, or advancing incitation per unit of information. Unwavering quality targets might incorporate upgrading sturdiness, diminishing the gamble of disappointment, or working on long-haul dependability.

- **Determine the Design Parameters:** Determine the design parameters and variables that have an impact on the actuator's effectiveness and dependability. Material properties, such as viscoelasticity and thermal properties; geometric factors, such as dimensions and shape; electrode configuration; actuation conditions, such as voltage and frequency; or control parameters are examples of these parameters.
- **Create a logical model:** Create a mathematical model of the twisted polymer actuator's behavior and how it relates to efficiency and dependability. The model ought to consider factors like energy dispersal, hysteresis, heat age, stress investigation, and some other applicable peculiarities. Solid mechanics, thermodynamics, electrical behavior, or a combination of these concepts can serve as the foundation for the model.
- **Create the function of the objective:** Make a goal capability that measures the ideal proficiency and unwavering quality results and characterizes the improvement issue. The earlier-described objectives, such as maximizing energy conversion efficiency or minimizing the likelihood of failure, should be reflected in the objective function. It fills in as the measurement to be improved during the planning cycle.
- **Choose a Method of Optimization:** Pick a fitting advancement strategy that suits the intricacy of the planned issue and addresses the proficiency and dependability goals. Hereditary calculations, molecule swarm improvement, angle-based strategies, or other advancement procedures can be utilized. Consider factors such as computational productivity, capacity to deal with requirements, and appropriateness for the particular issue.
- **Perform iterations of optimization:** Utilizing various combinations of design parameters, iteratively evaluate the objective function to carry out the optimization process. The optimization algorithm searches the design space for solutions that are either optimal or close to it. Every emphasis creates a bunch of plan factors that are assessed because of the goal capability and any pertinent requirements.
- **Assess and confirm solutions:** Assess the enhanced arrangements utilizing recreations, mathematical investigation, or exploratory testing. Using the established goals and pertinent performance metrics, evaluate the effectiveness and dependability of the optimized designs. To measure improvement, validate the optimized designs against desired specifications and compare them to initial or benchmark designs.

- **Evaluation of Robustness and Sensitivity:** To comprehend the impact of parameter variations and ensure that the optimized design is resilient to variations in operating conditions or material properties, conduct sensitivity analyses and robustness assessments. Consider the sensitivity of critical parameters that have a significant impact on reliability and efficiency during the design process.
- **Improvement by Iteration:** Emphasise the improvement interaction in light of the assessment results, awareness examination, and plan experiences. Adjust the design parameters, alter the objective function, investigate alternative optimization strategies, or take into account additional constraints to refine the design. The optimization iterations should be repeated until the design requirements are met or the efficiency and reliability goals are reached.
- **Approval and Execution Testing:** Approve the enhanced plan through thorough testing and execution assessment. Analyze the actuator's effectiveness, dependability, and stability over time using experiments. Compare the outcomes to the performance that was predicted by the optimization process. This step guarantees that the upgraded plan translates into further developed productivity and unwavering quality in useful applications.

The optimization design process can lead to twisted polymer actuators that are more reliable and efficient if these steps are followed as well as the relevant design parameters. It makes it possible for designers to find solutions that are either near-optimal or optimal in terms of energy efficiency, reliability, and meeting the application's specific needs.

6.3 Lessons Learned from Case Studies

The actual difficulties, design considerations, and possibility of making these actuators better. Case studies often teach us the following lessons:

- **Material Determination:** The performance of twisted polymer actuators is highly dependent on the polymer material selected. Case studies have demonstrated that choosing materials with the right viscoelastic properties, high actuation strains, and good mechanical stability is necessary to get the actuation performance you want.
- **Mathematical Plan:** Mathematical variables, like the shape, aspects, and perspective proportion of the actuator, fundamentally influence its exhibition. Contextual analyses have exhibited that streamlining the mathematical plan can improve activation range,

reaction time, and power yield. The fact that the actuator's stability and dependability are also affected by geometry emphasizes the significance of geometric considerations in the design process.

- **Configuration of the Electrode:** The game plan and positioning of cathodes assume a significant role in activation execution. Contextual analyses have shown that cautious anode configuration, including terminal arrangement, dividing, and math, can further develop activation qualities and limit undesired impacts like twisting or bowing. The actuator's behavior can be better controlled by optimizing the electrode configuration.
- **Methods of Production:** Twisted polymer actuators' performance and dependability are significantly influenced by the manufacturing procedure. Contextual investigations have highlighted the significance of exact manufacturing techniques, like controlled winding or arrangement, to accomplish predictable activation conduct. The final actuator's performance should be carefully considered when selecting fabrication methods.
- **Factors from the outside:** The presentation of turned polymer actuators can be impacted by natural circumstances like temperature and moisture. Case studies have shown that the actuator's behavior under various operating conditions must be taken into account during the design process. Appropriate portrayal and testing under significant ecological circumstances are pivotal for dependable actuator execution.
- **Mechanisms for Control and Driving:** To achieve actuation that is both precise and effective, twisted polymer actuators require specific control and driving mechanisms. Case studies have looked at voltage control, temperature control, and hybrid strategies for controlling things. The response time, stability, and energy efficiency of the actuator can all be improved by optimizing the control and driving mechanisms.
- **Stability and Longevity:** A major obstacle lies in ensuring that twisted polymer actuators are long-lasting and reliable. Long-term stability, resistance to fatigue, and maintaining actuation performance over extended usage have all been emphasized in case studies. Improving these actuators' dependability necessitates rigorous durability testing, careful consideration of the effects of the environment, and an understanding of failure mechanisms.
- **Optimization with Multiple Goals:** Multi-objective optimization strategies are useful for striking a balance between competing design goals, such as maximizing actuation range while reducing energy consumption, in case studies. Designers can investigate

trade-offs and identify optimal or near-optimal designs that satisfy specific application requirements by simultaneously considering multiple performance metrics.

- **Testimonial Support:** Exploratory approval is vital for surveying the exhibition of bent polymer actuators and checking the precision of plan advancements. Contextual analyses have underlined the significance of trial portrayal, benchmarking against starting plans or existing actuators, and approving the anticipated execution through testing. This iterative cycle refines the plan and upgrades the comprehension of the actuator's way of behaving.
- **Coordinated effort and interdisciplinary methodology:** In the design and optimization of twisted polymer actuators, case studies have demonstrated the advantages of collaboration and an interdisciplinary approach. The inclusion of specialists from fields like materials science, mechanics, electrical design, and control frameworks can prompt more thorough and inventive arrangements. Knowledge exchange, progress, and a greater comprehension of the actuator's performance are all facilitated by collaboration.

Researchers and designers can learn a lot about the development and optimization of twisted polymer actuators by looking at lessons learned from case studies.

Future Perspectives and Challenges

In any case, a few provocations should be tended to for the acknowledgment of these points of view. Here are a few future viewpoints and difficulties in turned polymer actuator research:

- **Enhanced Performance in Actuation:** Higher strain capabilities, quicker response times, and increased force output are some of the goals of future research aimed at enhancing the actuation capabilities of twisted polymer actuators. New materials, improved fabrication methods, and the optimization of actuator design parameters are all necessary for these enhancements to be realized. Improved actuation performance will also require recognizing and addressing issues like hysteresis, fatigue, and stress relaxation.
- **Multifunctionality and Joining:** The joining of various functionalities inside contorted polymer actuators is a promising future viewpoint. These actuators could include energy storage capabilities, self-healing properties, and sensing capabilities, according to research. This multifunctionality empowers the advancement of shrewd and independent frameworks with upgraded versatility and responsiveness. The creation and fabrication of integrated systems, the optimization of material combinations,

and ensuring compatibility between various functionalities are challenges in this regard.

- **Versatility and Assembling Methods:** It is still difficult to increase the production scale of twisted polymer actuators for use in real-world situations. The development of scalable manufacturing methods that can produce actuators with consistent performance on a larger scale is the primary focus of future research. This includes investigating cutting-edge techniques for additive manufacturing, roll-to-roll processing, and high-throughput fabrication. Beating difficulties connected with material choice, process control, and cost viability is vital for accomplishing the huge scope of the creation of turned polymer actuators.
- **Solidity and long-haul soundness:** The practical use of twisted polymer actuators necessitates ensuring their long-term stability and durability. Understanding and reducing degradation mechanisms, such as those caused by aging, fatigue, and environmental factors, will be the primary focus of subsequent research. It will be crucial to develop materials with improved mechanical and chemical stability and investigate novel strategies for increasing durability. Long-haul testing, sped-up maturing review, and prescient displaying can help with evaluating and working on the solidness of curved polymer actuators.
- **Control and integration of the system:** It is hard to coordinate the behavior of twisted polymer actuators and integrate them into larger systems. Control strategies that enable the precise and coordinated actuation of multiple actuators, resulting in complex and interactive systems, will be the primary focus of future research. Advanced feedback control algorithms, sensing methods, and communication protocols are all part of this. Control systems, electronics, and mechanical components must all be integrated for a system to be successfully integrated. This will necessitate interdisciplinary collaboration.
- **Biocompatibility and Biomedical Applications:** Biomedical applications like soft robotics, wearable devices, and biomedical implants could benefit from twisted polymer actuators. Future examinations will zero in on addressing difficulties connected with biocompatibility, biofouling, and long-haul solidity in organic conditions. Additionally, it will be essential to develop biocompatible and reliable interfaces between the actuators and biological systems. For biomedical applications to be successful, important considerations include ensuring safety, compatibility with physiological conditions, and regulatory compliance.

- **Energy Effectiveness and Manageability:** Further developing the energy effectiveness of curved polymer actuators and tending to their natural effect are significant future viewpoints. Scientists expect to upgrade actuator plans, foster energy-productive control methodologies, and investigate energy collection procedures to improve the general framework's effectiveness. In addition, the increasing emphasis on environmentally friendly and sustainable technologies coincides with the requirement for actuators to be biodegradable and recyclable.
- **Commercialization and Cost-Effectiveness:** Cost-effectiveness and commercial viability are crucial considerations for the widespread adoption of twisted polymer actuators. The pursuit of cost-effective materials, streamlining production methods, and cost-effective manufacturing process optimization are the primary areas of future research.

Emerging Trends in Twisted Polymer Actuator Research

- **High-level material Turn of events:** Analysts are effectively researching new polymer materials and composites with improved activation properties. This includes developing materials with novel functionalities, enhanced environmental stability, improved strain capabilities, and higher mechanical strength. Researchers hope to improve twisted polymer actuators' actuation range, dependability, and adaptability by investigating new materials.
- **Actuators With Several Responses:** Multi-responsive twisted polymer actuators that can be triggered by a variety of stimuli, including electric fields, temperature changes, light, humidity, and chemical triggers, are becoming increasingly popular. These multi-responsive actuators can enable complex actuation behaviors like shape memory effects, self-healing, or stimuli-triggered actuation and are adaptable to a variety of applications.
- **Applications of Soft Robots and Biomimetic Systems:** Turned polymer actuators are tracking down expanding applications in the field of delicate advanced mechanics and biomimetics. Analysts are investigating the incorporation of curved polymer actuators into delicate mechanical frameworks, wearable gadgets, and biomimetic designs to imitate normal developments, accomplish able control, and improve human-robot connection. The development of actuators with enhanced biocompatibility, flexibility, and bioinspired designs is being driven by this trend.
- **Self-powered and energy-harvesting systems:** The idea of using twisted polymer actuators' actuation capabilities for energy

harvesting is getting more and more attention. Scientists are investigating the capability of changing mechanical misshapen Ings during activation into electrical energy, consequently empowering self-controlled frameworks. By utilizing the inherent mechanical-to-electrical energy conversion capabilities of actuators, this trend aims to improve devices' energy efficiency and autonomy.

- **Producing Advances:** The improvement of cutting-edge fabricating methods and cycles is critical for understanding the maximum capacity of curved polymer actuators. To produce intricate actuator designs with high precision, repeatability, and scalability, researchers are looking into additive manufacturing (also known as 3D printing), microfabrication, and scalable manufacturing techniques. These advances work with the creation of actuators with customized calculations, anode arrangements, and joining with different parts.
- **Simulation and modeling:** Understanding the behavior of twisted polymer actuators and improving their performance are greatly aided by modeling and simulation. To accurately predict the actuation response, optimize design parameters, and simulate the actuator's behavior under various operating conditions, advanced mathematical models, computational tools, and simulation techniques are being developed by researchers. This pattern empowers more effective and informed planning processes.
- **Brilliant Control and Detection:** An emerging trend involves combining twisted polymer actuators with smart control and sensing technologies. Analysts are investigating control frameworks, AI calculations, and implanted sensors to upgrade the actuator's exhibition, flexibility, and independent activity. For improved precision and response, this includes closed-loop control, intelligent actuation strategies, and real-time actuator behavior monitoring.
- **Biomedical and medical services Applications:** Twisted polymer actuators have the potential to be used in a variety of biomedical and healthcare settings. In areas like soft prosthetics, surgical robotics, drug delivery systems, microfluidics, and wearable medical devices, their application is being investigated by researchers. The biocompatibility, controllability, and consistency of these actuators make them reasonable for biomedical applications, where they can empower exact and negligibly intrusive intercessions.
- **Sustainability of the environment:** The improvement of naturally reasonable wound polymer actuators is acquiring significance. To reduce the actuators' negative impact on the environment, researchers are looking into bio-based and biodegradable polymers, eco-friendly manufacturing processes, and recyclable materials.

Key Challenges and Limitations

- **Limits of the strain:** Contorted polymer actuators commonly show lower strain capacities compared with different kinds of actuators, for example, shape memory combinations or electroactive polymers. It is still difficult to achieve higher actuation strains without sacrificing other performance metrics. A major objective of the research is to increase the strain output while maintaining mechanical stability and long-term durability.
- **Hysteresis and Reaction Time:** Wound polymer actuators frequently show hysteresis, where the activation reaction occurs during stacking and isn't entirely reversible during dump cycles. This hysteresis can restrict the exactness and repeatability of the invitation. Twisted polymer actuators' limited use in high-speed actuation systems is also due to their relatively slow response times. Areas of active research include overcoming hysteresis and increasing response time.
- **Fatigue and mechanical stability:** When subjected to significant deformations, twisted polymer actuators may exhibit mechanical instability, such as buckling or curling. For dependable and predictable performance, it is essential to ensure mechanical stability throughout the actuation cycle and under various loading conditions. Because repeated actuation cycles can result in material degradation and decreased performance over time, fatigue life and long-term stability are also important considerations.
- **Control and Detecting:** Accomplishing exact and powerful control of bent polymer actuators is testing. Accurate and repeatable actuation requires the development of efficient control algorithms and sensing methods for real-time monitoring and regulation of actuator behavior. Research into the development of smart control strategies and the integration of sensors for feedback control is ongoing.
- **Ecological Awareness:** Temperature, humidity, and chemical exposure can all have an impact on twisted polymer actuators. The stability and performance of the actuation can be affected by changes in these environmental conditions. It is essential for practical applications to devise methods for reducing the environmental sensitivity of twisted polymer actuators and guaranteeing their consistent operation under a variety of operating conditions.
- **The complexity of manufacturing and scalability:** increasing the manufacture of wound polymer actuators while keeping up with predictable execution is a test. The assembling system can be

perplexing and requires exact control of material properties, math, and cathode design. Creating versatile assembling procedures that can deliver actuators with solid and reproducible execution is fundamental for down-to-earth execution.

- **Stability and Longevity:** For their practical applications, it is essential to ensure that twisted polymer actuators are long-lasting and reliable. The actuators should be able to withstand mechanical and environmental stresses and maintain their performance over long periods of use. These actuators can be made more reliable and durable by improving their fatigue resistance, understanding failure mechanisms, and conducting rigorous durability testing.
- **Cost and viability in business:** The integration of twisted polymer actuators, as well as the price of the materials used in the fabrication process, may prevent widespread adoption. Research endeavors are expected to improve the cost adequacy of these actuators through productive assembling techniques, material determination, and interaction computerization. Accomplishing business feasibility and versatility will be fundamental for their integration into different ventures and applications.

Materials scientists, mechanical engineers, experts in control systems, and other relevant fields must collaborate across disciplines to address these issues and limitations. Proceeding with examination and development is important to beat these obstacles and unlock the maximum capacity of contorted polymer actuators.

Potential Solutions and Future Directions

Researchers are actively looking into potential solutions and future directions to address the difficulties and limitations of twisted polymer actuators. The following are some important areas:

- **Material Progress:** Scientists are examining new polymer materials and composites with further developed activation properties, for example, higher strain capacities, quicker reaction times, and improved mechanical soundness. This includes fitting the material piece, atomic design, and handling methods to accomplish the desired incitation execution. It is anticipated that breakthroughs in nanotechnology and material science will result in the creation of novel materials with superior actuation properties.
- **Half-and-half actuator frameworks:** The individual limitations of twisted polymer actuators can be overcome by combining them with other actuation technologies. The overall actuation performance of hybrid actuator systems can be improved by

combining twisted polymer actuators with shape-memory alloys, electroactive polymers, or other types of actuators. Researchers hope to achieve enhanced strain capabilities, quicker response times, and enhanced reliability by synergistically utilizing the strengths of various actuation mechanisms.

- **Advanced Methods of Control:** For robust and precise actuation control, it is essential to develop advanced control strategies. Specialists are investigating model-based control calculations, versatile control approaches, and AI strategies to upgrade incitation execution, relieve hysteresis, and further develop reaction time. Closed-loop control and precise manipulation of twisted polymer actuators are made possible by real-time feedback control systems that are integrated with precise sensing technologies.
- **Self-Detecting Actuators:** The twisted polymer actuators' sensing capabilities can be directly integrated into them, which can simplify the control system and boost actuation performance. To monitor actuator deformation and strain, researchers are looking into self-sensing mechanisms like embedding strain sensors or utilizing the polymer's inherent electrical conductivity. Real-time feedback control without the use of external sensors is made possible by this self-sensing capability, which improves actuation precision and dependability.
- **Simulation and computational modeling:** Understanding the behavior of twisted polymer actuators and optimizing their performance require mathematical modeling and simulation. Scientists are creating advanced computational models that think about material properties, mathematical setups, and outside improvements. Virtual prototyping, design optimization, and actuator behavior prediction under various operating conditions are all made possible by these models. The design and optimization of twisted polymer actuators will be made easier by increasing the models' accuracy and fidelity.
- **Miniature and Nanofabrication Methods:** Twisted polymer actuators can now be precisely controlled thanks to advancements in micro- and nanofabrication methods. To create actuators with customized geometries, hierarchical structures, and intricate electrode configurations, researchers are investigating methods like 3D printing, microfabrication, and self-assembly. These methods empower the creation of actuators with improved execution, dependability, and reconciliation abilities.
- **Designs that are inspired by nature:** Researchers are looking into biomimetic designs for twisted polymer actuators, taking their

cues from the natural world. Achieving actuators with enhanced adaptability, efficiency, and performance can be accomplished by imitating natural structures and mechanisms. For instance, bio-propelled plans because of the standards of muscle-like designs or plant-motivated developments can improve the activation capacities of curved polymer actuators.

- **Normalization and Commercialization:** Twisted polymer actuators' commercialization and widespread adoption will be aided by the establishment of standardized manufacturing guidelines, performance metrics, and testing protocols. To establish quality control and establish industry standards, academic institutions, businesses, and regulatory bodies must work together. Twisted polymer actuators' commercial viability will also be driven by advancements in cost-effective production methods and scalable manufacturing processes.

By tending to these arrangements and investigating future headings, specialists mean to defeat the difficulties and restrictions of curved polymer actuators, opening their true capacity for a great many applications in mechanical technology, medical care, aviation, and then some. These actuators can be put into use and incorporated into real-world systems if more research and innovation are conducted.

CONCLUSION

They introduce a top-to-bottom investigation of bent polymer actuators and their streamlining plan in the field of strong mechanics. The study's introduction focused on twisted polymer actuation's distinctive features and potential applications. The foundation and inspiration driving the exploration were examined, stressing the requirement for improved activation execution, productivity, and dependability in the plans of these actuators.

The research's goals were laid out, which included figuring out what factors, difficulties, and design parameters affect how well-twisted polymer actuators work. To achieve optimal actuation performance, the significance of material selection and characterization, performance evaluation and testing, mathematical modeling and simulation, and optimization techniques were emphasized.

Twisted polymer actuators' potential in a variety of fields, including soft robotics, biomedical devices, and adaptive structures, was demonstrated through case studies of successful applications. These case studies' design considerations and difficulties were discussed, drawing attention to the significance of interdisciplinary collaboration and addressing specific application requirements.

The exploration further dug into the streamlining plan of bent polymer actuators, zeroing in on improving activation execution, effectiveness, and unwavering quality. The utilization of multi-objective improvement draws near; hereditary calculations and developmental strategies were investigated as successful procedures for accomplishing ideal plans. Also, responsiveness investigations and parametric examinations were talked about to figure out the impact of plan boundaries on actuator execution.

The need for suitable materials with desirable mechanical properties and environmental stability was also emphasized in the chapter's discussion of the significance of material selection and characterization. Methods for testing and performance evaluation were discussed to guarantee an accurate and reliable evaluation of the actuator's performance.

The potential applications of mathematical modeling and simulation techniques for comprehending the behavior of actuators, forecasting performance, and optimizing designs were investigated. The utilization of computational models empowered virtual prototyping and plan improvement, giving significant bits of knowledge into actuator conduct and directing the planning cycle.

The study brought to light some of the newest trends in the study of twisted polymer actuators. These include the investigation of biocompatible materials and biomedical applications as well as the integration of multiple functionalities like sensing and self-healing. The ongoing efforts to enhance the capabilities and potential applications of twisted polymer actuators are reflected in these trends.

The key difficulties and limitations of twisted polymer actuators, such as strain limitations, hysteresis, mechanical stability, environmental sensitivity, and scalability, were finally addressed. It was emphasized that future research is required to overcome these obstacles and improve actuator designs.

The developments in a variety of fields could greatly benefit from twisted polymer actuators. Through an improvement plan, material choice, execution assessment, and numerical display, scientists can upgrade incitation execution, proficiency, and unwavering quality. The practical implementation and integration of twisted polymer actuators into real-world systems, which will contribute to advancements in solid mechanics and enable new applications in robotics, healthcare, aerospace, and beyond, will be made possible by addressing the challenges and limitations and examining emerging trends.

5

Elasticity and Mechanical Testing of Solid Mechanics

INTRODUCTION

The study of the behavior of solid materials when they are subjected to external forces and deformations are the primary focus of the solid mechanics subfield of engineering mechanics. For various engineering applications, designing structures that can withstand loads and stresses requires an understanding of the mechanical properties of solid materials. One essential property of materials in strong mechanics is flexibility, which alludes to their capacity to get back to their unique shape following being disfigured by an applied force. Material characterization, structural analysis, and design optimization all depend on the study of elasticity and mechanical testing of solid materials.

The field of flexibility traces back to the seventeenth century when Robert Hooke figured out Hooke's Regulation, which expresses that the disfigurement of a versatile material is straightforwardly relative to the applied pressure inside as far as possible. Hooke's Regulation gives a basic system for figuring out the direct, versatile way of behaving with materials, where anxiety is directly related. However, to accurately describe and characterize the mechanical properties of materials that exhibit more complex behavior beyond the elastic limit, such as plastic deformation, viscoelasticity, or nonlinear response, more advanced constitutive models and testing methods are required.

The purpose of mechanically testing solid materials is to measure how well they respond to external forces and deformations. This information is useful for choosing materials, making designs work better, and analyzing structures. Mechanical tests are performed under controlled conditions to assess different mechanical properties like versatile modulus, yield strength, extreme strength, pliability, durability, and weakness opposition. These properties are essential for figuring out the material's behavior and anticipating its conduct in certifiable applications.

Malleable testing is one of the most well-known mechanical tests used to determine the mechanical properties of materials under pressure. A specimen is subjected to an axial load and the deformation and stress that result are measured. In contrast, compression testing involves placing a compressive load on the material. A material's behavior under bending loads is evaluated during flexural testing, while its response to shear forces is measured during shear testing. Impact testing evaluates a material's capacity to absorb energy under sudden loading conditions, while hardness testing reveals a material's resistance to indentation or scratching. Exhaustion testing assesses the material's protection from cyclic stacking, while creep testing looks at its twisting under a consistent burden over a drawn-out period.

The mechanical testing of strong materials requires exact instrumentation and government-sanctioned testing techniques to guarantee precise and solid outcomes. Current testing contraptions incorporate load cells, extensometers, strain measures, and relocation sensors that catch the mechanical reaction of materials during testing. Material properties can be extracted, and the mechanical behavior of the material can be characterized using advanced analysis methods and data acquisition systems.

Mechanical testing can be used for a wide range of purposes in a variety of industries to characterize materials. In primary design, the mechanical properties of materials are fundamental for planning protected and productive structures, spans, and other foundations. Materials' suitability for use in critical parts subjected to extreme conditions is determined by their mechanical behavior in the automotive and aerospace industries. Mechanical testing is essential for assessing the performance and safety of medical implants and prosthetic devices in biomechanics and biomaterials. Additionally, understanding the behavior of new materials at various scales and under various loading conditions necessitates mechanical testing, which plays a crucial role in material science research.

Progressions in testing methods and hardware, as well as the joining of computational display and re-enactment, have extended the abilities of mechanical testing as of late. Nanoindentation and microscale testing empower the portrayal of material properties at small scopes, giving bits of knowledge into the mechanical way of behaving of slim films, coatings, and microstructures. Full-field deformation analysis is made possible by digital image correlation (DIC) techniques' non-contact measurement of surface displacements and strains. In-situ mechanical testing considers the perception of material conduct under constant stacking conditions utilizing progressed microscopy and imaging strategies.

Background and Significance

The study of how solid materials react to external forces and deformations is the focus of the engineering mechanics subfield known as solid mechanics. Structural engineering, materials science, aerospace engineering, biomechanics, and numerous other engineering fields all rely heavily on it. Understanding the mechanical properties of strong materials is fundamental for planning protected and productive designs, anticipating their presentation under various stacking conditions, and advancing their plan to meet explicit necessities.

In solid mechanics, elasticity is a fundamental property of materials that describes their capacity to revert to their original shape after being deformed by an external force. This conduct is normally seen inside the versatile furthest reaches of a material, past which extremely durable deformity or disappointment might happen. Flexibility is represented by Hooke's Regulation, which expresses that the misshapen Ing of material is straightforwardly relative to the applied pressure inside the versatile reach. Hooke's Regulation gives a straightforward direct connection among anxiety, permitting specialists to break down and foresee the way of behaving of flexible materials under various stacking conditions.

Due to its numerous applications, the study of elasticity is crucial to solid mechanics. In structural engineering, designing load-bearing structures like bridges, buildings, and dams requires an understanding of material elasticity. Engineers can accurately predict the structural response to various loads by taking into account the elastic behavior of materials. This keeps the structures within safe deformation and stress limits. Flexibility is likewise imperative in the plan of mechanical parts and hardware, where the appropriate choice of materials with explicit versatile properties guarantees ideal execution and life span.

Besides, versatility assumes a fundamental part in materials science and examination. By concentrating on the flexible way of behaving materials, specialists can acquire bits of knowledge about their microstructural properties, like the game plan and communication of particles or atoms inside the material. This information helps in grasping material deformities, mechanical security, and the connection between design and properties. Additionally, elasticity plays a crucial role in the creation of novel materials and composites with specialized properties for particular applications. By controlling the versatile way of behaving, specialists can plan materials with upgraded strength, adaptability, and protection from the deformity.

The mechanical testing of solid materials is essential for defining their mechanical properties and behavior, in addition to elasticity. Mechanical testing considers the estimation of boundaries like flexible modulus, yield strength, extreme strength, pliability, sturdiness, and exhaustion obstruction. The ability of a material to withstand external loads, its response to deformation, and its resistance to failure can all be learned from these properties. Engineering design, material selection, quality control, and failure analysis all depend on this kind of data.

Mechanical testing techniques that are both accurate and effective have become increasingly important as a result of advancements in materials science and engineering. To accurately measure mechanical properties at various length scales, from macroscale structures to microscale components, engineers and researchers constantly strive to develop new testing methods and equipment. To evaluate material properties without harming the specimen, research has also been focused on the development of non-destructive testing methods like digital imaging and ultrasound.

Understanding the foundation and meaning of versatility and mechanical testing in strong mechanics gives an establishment to additional investigation and exploration in this field. Engineers can make well-informed choices regarding the selection of materials, design optimization, and structural analysis by studying and characterizing the mechanical properties of solid materials. This will result in the creation of engineering solutions that are safer, more effective, and novel.

Objectives

The following are the research objectives for elasticity and mechanical testing in solid mechanics:

1. To give a complete comprehension of the hypothetical standards of fundamental versatility and mechanical conduct in strong materials. To lay a strong foundation for future research, this entails investigating fundamental ideas like stress, strain, and constitutive equations.
2. To audit and dissect different mechanical testing methods used to assess the mechanical properties of strong materials. This incorporates tractable testing, pressure testing, flexural testing, shear testing, hardness testing, influence testing, weakness testing, and creep testing. The goal is to figure out the standards, benefits, impediments, and utilisations of each testing technique.
3. To investigate the mechanical testing standards, procedures, and experimental equipment. This includes examining the instrumentation and hardware expected for precise and dependable testing, test planning procedures, and adherence to laid-out testing conventions and guidelines.

4. To investigate the characterization of mechanical properties like toughness, fracture toughness, residual stress analysis, elastic modulus, Poisson's ratio, yield strength, and ultimate strength. Understanding these properties' significance in material selection, design optimization, and structural analysis is the goal.
5. To investigate cutting-edge mechanical testing methods like nanoindentation, microscale testing, digital image correlation (DIC), in-situ mechanical testing, computational modeling, and non-destructive evaluation methods. The goal is to grasp the abilities and limits of these strategies and their possible applications in material portrayal and primary examination.
6. To address the difficulties and limitations of mechanical and elasticity testing, such as measuring accurately, testing in extreme conditions, characterization at multiple scales and physics, integrating experimental and computational methods, and standardization and certification. The objective is to suggest potential solutions to these difficulties and identify areas for improvement.
7. To feature the meaning of flexibility and mechanical testing in designing applications, including primary designing, materials science, aviation design, biomechanics, and others. The goal is to emphasize how important it is to understand and characterize the mechanical properties of solid materials in real life and their practical implications.
8. To identify perspectives for the future and research directions in the fields of mechanical testing and elasticity. The goal is to investigate emerging patterns, advancements, and areas of examination that can additionally propel the comprehension and portrayal of mechanical conduct in strong materials.

By and large, the goals of the exploration are to give an exhaustive survey of versatility and mechanical testing in strong mechanics, advance a more profound comprehension of material ways of behaving, and contribute to the improvement of exact and productive testing techniques that can upgrade material portrayal and primary examination in different designing applications.

ELASTICITY THEORY

Overview of Elasticity

Versatility is a key idea in strong mechanics that portrays the capacity of a material to misshape reversibly when exposed to an outside power and return to its unique shape upon the expulsion of the power. It's a property that lets materials deform temporarily when loads are applied without changing their shape or structure forever. To comprehend and anticipate the mechanical behavior of materials, the study of elasticity is essential for a variety of engineering applications.

At the point when a material is exposed to an outside force, it encounters inward burdens that cause twisting. When a material can store this energy and regain its original shape when the force is removed, this is called elastic deformation. Stress and strain are indicators of how a material reacts to an external force.

The internal forces exerted on material by external forces are measured by stress. The abbreviation for it is "," and its definition is "force per unit area." There are various types of stress, such as tensile stress, which is caused by forces pulling the material apart; compressive stress, which is caused by forces pushing the material together; and shear stress, which is caused by forces that are parallel to the surface of the material.

In contrast, strain is a measurement of the material's deformation as a result of the applied stress. It is characterized as the proportion of the adjustment of the length or state of a material to its unique length or shape and is signified by the impalement of the length or state of a material to its unique length or shape and is signified by the image. Strain is additionally arranged into various kinds, including malleable strain (prolongation of the material), compressive strain (shortening of the material), and shear strain (contortion of the material).

The connection among anxiety is represented by Hooke's Regulation, which expresses that, inside the versatile furthest reaches of a material, the pressure is straightforwardly corresponding to the strain.

The material's stiffness, or resistance to deformation, is measured by its elastic modulus. Within the elastic region, it depicts the slope of the stress-strain curve.

The stress-strain curve, which depicts the material's response to applied stress, is frequently used to describe elasticity. Following Hooke's Law, the material initially behaves linearly and elastically at low-stress levels. The material may reach its elastic limit as applied stress increases, at which point it undergoes plastic deformation or permanent shape changes. The yield point is the starting point for plastic deformation. Assuming the pressure is eliminated after plastic deformity, the material will show long-lasting deformity or may try and come up short.

The versatile way materials behave can be isotropic or anisotropic. Isotropic materials show similar mechanical properties this way and that, though anisotropic materials have various properties every which way. Anisotropy emerges from the primary game plan and creation of the material, like the arrangement of gems or filaments.

Understanding the standards of flexibility is essential for architects and researchers associated with material determination, foundational layout, and examination. By precisely describing the versatile way materials behave, architects can anticipate and control misshapen Ings, guarantee primary respectability, and enhance plans to meet explicit execution necessities. Besides, flexibility shapes the reason for additional complicated material models that record nonlinear behavior, viscoelasticity, and time-subordinate disfigurement, empowering a more thorough comprehension of material reactions under different stacking conditions.

Stress and Strain

In strong mechanics, anxiety is a central idea that portrays the mechanical way materials behave under outer powers. They assume an essential role in figuring out the misshapen Ing and disappointment of materials, as well as in the planning and examination of designs and parts.

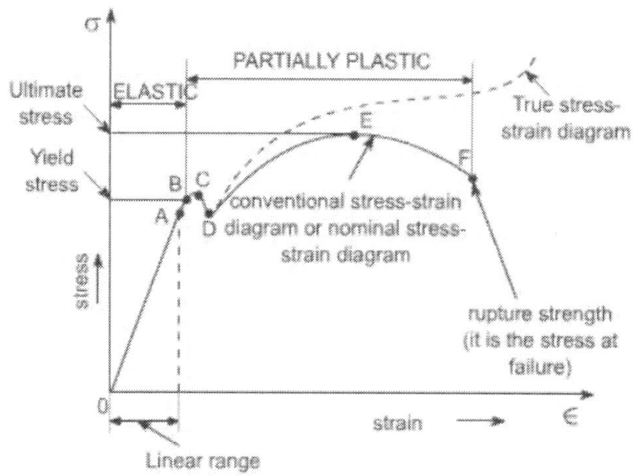

Figure 1: A typical test curve for the mid steel

A material's internal force per unit area caused by external forces is referred to as stress. It is a proportion of the power of the powers following up on the material. Stress is commonly indicated by the image and has units of power per unit region (e.g., Pascal or N/m2). The extent and conveyance of stress inside a material impact its misshapen Ing and reaction to outside loads.

In contrast, strain is a measurement of a material's deformation or change in shape under stress. It measures the relative change in the size or state of a material because of applied power. The symbol denotes strain, which has no dimensions. It is frequently expressed as a percentage or ratio. The process by which a material expands, contracts, or distorts in response to stress is referred to as strain.

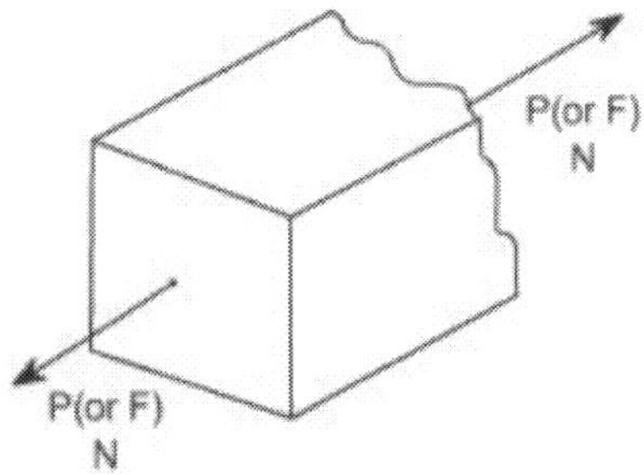

Figure 2: Stress

A stress-strain curve, which shows how the material behaves under various loading conditions, typically depicts the relationship between stress and strain. According to Hooke's Law, stress and strain are inversely proportional to each other in the curve's elastic region. This straight relationship is portrayed by the flexible modulus (E), otherwise called Young's modulus. The versatile modulus addresses the firmness of a material and characterizes its protection from distortion. Within the elastic limit, it is defined as the ratio of strain to stress:

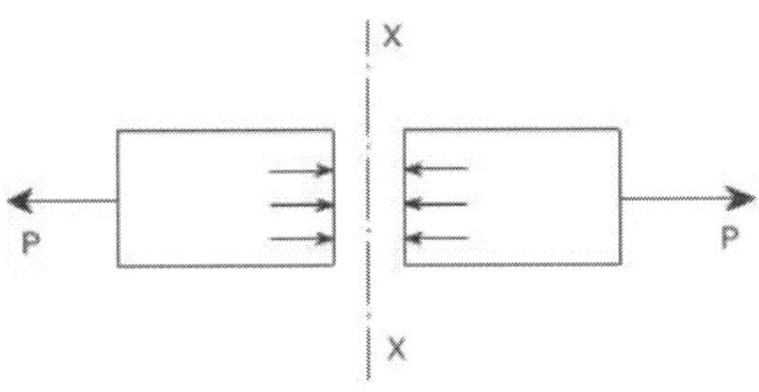

Figure 3: Flexible Modulus

Various materials have different versatile moduli, reflecting their innate firmness or adaptability. Under applied loads, materials with high elastic moduli, like steel, deform only slightly, whereas materials with low elastic moduli, like rubber, deform significantly.

Materials may exhibit nonlinear behavior when the stress-strain relationship deviates from linearity beyond the elastic limit. Plastic deformation, yielding, or failure can cause this. The stress-strain curve may either exhibit distinct regions of behavior, such as strain hardening or strain softening, or it may exhibit a yield point, which signifies the beginning of plastic deformation.

In structural analysis, material characterization, and design optimization, stress and strain are crucial parameters. Stress and strain calculations are used by engineers and scientists to assess structures' strength and stability, forecast material failure, and determine which materials are best suited for particular applications. Understanding the anxiety conveyance inside a material empowers the recognizable proof of expected flimsy parts, the assessment of primary trustworthiness, and the streamlining of plans to upgrade execution and security.

They give insight into the way materials behave under various stacking conditions and are fundamental for planning and examining designs and parts.

Constitutive Equations

Mathematical relationships that describe the mechanical behavior of materials under various loading conditions are known as constitutive equations. They link stress and strain, making it possible for scientists and engineers to study and predict how materials will react to applied forces.

The particular material properties and the kind of deformation are considered to determine the constitutive equations. Various materials show different mechanical ways of behaving, going from straight versatile to exceptionally nonlinear and time-subordinate reactions. The determination of constitutive condition relies upon the material's attributes and the degree of exactness expected for the examination.

A few common constitutive equations used in solid mechanics will be discussed here:

Hooke's Regulation: The linear elastic behavior of materials within their elastic limit is described by Hooke's Law, which is a fundamental constitutive equation. It expresses that the pressure is straightforwardly relative to the strain inside this reach: = E, where E is the material's stiffness and Young's modulus is its elastic modulus. The material is assumed to be isotropic and to have a linear relationship between stress and strain according to Hooke's Law.

Elasticity is not linear; materials frequently exhibit non-linear elastic behavior. Nonlinear flexible constitutive conditions catch the nonlinear connection among anxiety. Nonlinear effects like strain hardening and softening can be accounted for by these equations. The hyperplastic material model, which describes the behavior of elastomers, rubber-like materials, and biological tissues, is one model that is frequently utilized.

Plasticity: Versatility models are utilized to portray the extremely durable or irreversible distortion of materials past their flexible breaking point. Plasticity models take into account yielding, strain hardening, and material flow under stress. The Von Mises yield criterion, which describes the onset of yielding in ductile materials, is the most well-known plasticity model.

Viscoelasticity: Both the elastic and the time-dependent (viscous) behavior of materials can be explained by using viscoelastic constitutive equations. Materials like polymers and biological tissues that exhibit time-dependent deformation under applied stress are suitable candidates for these models. The Kelvin-Voigt model and the Maxwell model are two examples of models that combine elastic and viscous components to describe behavior.

Creep: Under constant load or stress, creep is the time-dependent deformation of a material. Creep constitutive conditions capture this way of behaving and are generally utilized in high-temperature applications or materials exposed to long-haul stacking. The power-regulation killjoy condition is a broadly utilized model that relates the strain rate to pressure and temperature.

Fracture and Damage: Damage and fracture mechanics can also be used in constitutive equations to describe how materials behave under high stresses or repeated loading. To predict a material's failure, these models take crack propagation, energy dissipation, and failure criteria into account.

It is essential to keep in mind that constitutive equations are typically derived using theoretical assumptions and experimental data. The level of complexity taken into consideration and the particular material behavior being modeled both affect these equations' precision. High-level constitutive models might require boundaries not entirely set in stone through trial testing or aligned because of exact information.

In general, the foundation for analyzing the mechanical behavior of materials is constitutive equations. Engineers and scientists can accurately predict material response, simulate structural behavior, and optimize designs for a variety of applications by selecting an appropriate constitutive equation and determining the material parameters.

Linear Elasticity and Hooke's Law

Straight versatility is a crucial idea in strong mechanics that depicts the way materials behave inside their flexible breaking point, where the pressure-strain relationship keeps Hooke's Regulation. It provides a better at this point significant guess for the majority of design applications.

Hooke's Regulation expresses that the pressure in a material is straightforwardly corresponding to the strain inside the direct versatile reach. The material's stiffness and resistance to deformation are measured by its elastic modulus.

The material exhibits linear stress-strain relationships in its linear elasticity regardless of the magnitude of the applied load. This indicates that, when subjected to light loads, the material deforms elastically before reverting to its original form. The material's structure is not permanently altered, and the deformation can be reversed.

The flexible modulus, E, is a material property that measures its firmness. The ratio of stress to strain is how it is defined:

The elastic modulus is unique to each material and is typically expressed in units of stress (such as Pa or psi). The inherent mechanical properties of various materials are reflected in the various elastic moduli of those materials.

Under relatively light loads, materials like metals, ceramics, and some polymers exhibit linear elastic behavior. Nonetheless, it is essential to take note that not all materials show direct versatility. Even within the elastic limit, some materials, like rubbers and elastomers, exhibit nonlinear elastic behavior, in which the relationship between stress and strain deviates from a straight line. The material behavior is described by more intricate constitutive models like hyper elasticity in these instances.

Straight versatility and Hooke's Regulation give a rearranged, valuable structure at this point for figuring out the twisting and reaction of materials under little loads. Numerous engineering calculations, including structural analysis, material selection, and design optimization, are based on them. Engineers can accurately predict and control the deformation of structures, guarantee their integrity, and optimize designs for maximum performance and safety by comprehending the linear elastic behavior of materials and the elastic modulus.

Anisotropic Elasticity

The mechanical way of behaving of materials that show various properties in every way. As opposed to isotropic materials, which have similar mechanical properties this way, anisotropic materials have particular material properties along various tomahawks or directions.

The internal structure and arrangement of the material's constituents, such as crystals, fibers, or layers, is what causes the anisotropic behavior. Because of their microstructure's anisotropy, these materials naturally depend on one another in different directions.

A tensorial form is used to describe the stress-strain relationship in anisotropic elasticity, where elastic constants that change with direction link the stress and strain components. Depending on the material's complexity and symmetry, various coordinate systems, such as Cartesian or tensor notation, are typically used to describe the anisotropic behavior.

The degree of anisotropy of an isotropic material can vary. Some materials may behave orthotopically, with distinct properties along each of their three symmetrically perpendicular axes. Transverse isotropy, in which a material has one preferred direction and properties that are identical in all other directions perpendicular to that axis, may occur in other materials. Fiber-reinforced composites, for example, can exhibit full anisotropy, in which their mechanical properties change continuously with orientation.

To portray and investigate anisotropic materials, different constitutive models are utilized. The anisotropic behavior is described by these models using direction-dependent elastic constants like the elastic modulus, shear modulus, and Poisson's ratio. In typical methods, the material properties and their transformations concerning various coordinate systems are typically represented utilizing stiffness matrices or tensors.

Applications in engineering are significantly impacted by materials' anisotropic behavior. Structures and parts made of anisotropic materials are affected in design, analysis, and performance. When selecting materials, predicting their behavior under various loading conditions, and optimizing designs to ensure the desired performance and safety, engineers must carefully consider their anisotropic properties.

In general, anisotropic elasticity is a crucial idea in solid mechanics that describes the mechanical behavior of materials with properties that change in response to direction. Grasping anisotropic materials and precisely demonstrating their way of behaving is vital for different design disciplines, including aviation, auto, structural design, and materials science.

Nonlinear Elasticity

The mechanical behavior of materials with a nonlinear relationship between stress and strain, in contrast to the linear relationship described by Hooke's Law, is referred to as "nonlinear elasticity." In contrast to directly versatile materials, nonlinearly flexible materials don't have a consistent versatile modulus and can show an extensive variety of complicated disfigurement reactions under applied loads.

Large deformations, nonlinear material properties, strain rate effects, temperature dependence, and microstructural changes are all potential causes of nonlinear behavior. These materials might show, for example, strain solidifying, strain relaxing, hysteresis, and nonlinear pressure strain bends.

A few constitutive models have been created to depict nonlinear versatility, contingent upon the particular material way of behaving and application. Several common models include:

Hyperplastic materials are a sort of nonlinearly versatile material that shows a strain-energy thickness capability. Elastomers, rubbers, and biological tissues are often described using these materials. The material's response to deformation can be captured by the strain-energy density function, which can be obtained from experimental data or theoretical models like the Mooney-Rivlin, Ogden, or Neo-Hookean models.

- **Plasticity:** Pliancy models are utilized to portray the long-lasting disfigurement and plastic way of behaving of materials past their flexible cut-off. Plastic deformation occurs when the material undergoes shape changes that cannot be reversed and the applied stress exceeds a yield point. Versatility models consider factors such as yielding, strain solidifying, and stream rules to portray the material's conduct under plastic distortion. The Von Mises yield measure and related stream rules are regularly utilized in pliancy displaying.
- **Viscoelasticity:** Viscoelastic materials combine elastic and viscous responses to exhibit time-dependent behavior. When put under stress, these materials exhibit both immediate elastic deformation and delayed viscous relaxation. Polymers and biological tissues frequently exhibit viscoelastic behavior. To describe the time-dependent deformation of viscoelastic materials, models like the Maxwell, Kelvin-Voigt, and Prony series are utilized.
- **Mechanics of Nonlinear Fractures:** Models of nonlinear fracture mechanics focus on crack propagation and fracture and how materials behave under extreme stress. Concepts like crack tip plasticity, damage mechanics, and fracture toughness are incorporated into these models to account for crack initiation, propagation, and energy dissipation.

Numerous engineering applications, such as structural analysis, material characterization, and design optimization, rely heavily on nonlinear elasticity. Predicting the response of materials to complex loading conditions, ensuring the structural integrity of components, and optimizing designs for improved performance and safety all require a thorough understanding of and accurate modeling of the nonlinear behavior of materials. It is essential to note that sophisticated experimental testing, material characterization methods, and numerical simulations are required to capture the nonlinear behavior of materials. The decision of the suitable constitutive model relies upon the particular material properties, stacking conditions, and degree of exactness desired for the investigation.

MECHANICAL TESTING TECHNIQUES

Tensile Testing

It is a fundamental test that provides useful information regarding the ductility, elasticity, and strength of a material. In engineering, manufacturing, materials science, quality control, and other fields, tensile testing is frequently used.

The essential target of elastic testing is to apply a progressively expanding uniaxial burden to a test example until it breaks. The material's response to tension, including its stress-strain behavior, ultimate tensile strength, yield strength, elongation, and modulus of elasticity, can be measured as a result of this.

A universal testing machine (UTM) with specialized grips for securely holding the test specimen is typically used for the tensile test. The test example, frequently a normalized round and hollow shape called a dogbane or hand weight, is ready as indicated by unambiguous guidelines, like ASTM or ISO.

During the malleable test, the test example is clasped between the holds of the UTM, guaranteeing that it stays adjusted along the stacking pivot. The specimen is stretched and elongated until it eventually breaks apart as a result of the tensile force applied by the machine. All through the test, the UTM constantly records the heap applied and the comparing deformity.

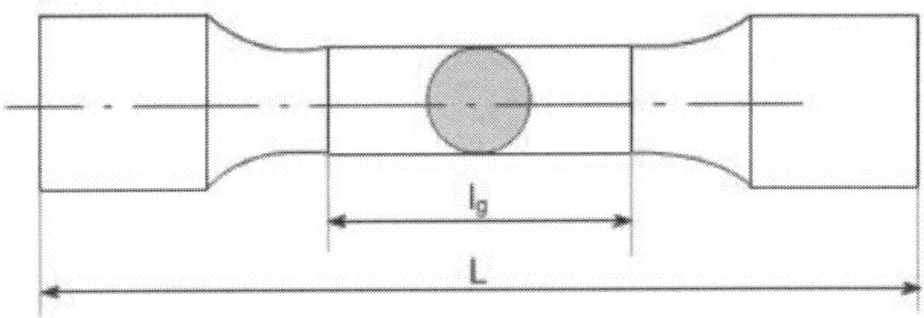

Figure 4: utilizes a circular X-section

- **Stress-Strain Bend:** The material's response to tension is depicted graphically on the stress-strain curve. Throughout the test, it demonstrates the connection between stress and strain. The bend commonly displays various areas, including the versatile locale, yielding or plastic distortion district, and crack locale.
- **Ultimate Strength (UTS):** The maximum stress a material can withstand before breaking is its ultimate tensile strength. It is an important parameter for material selection and design because it shows the material's maximum resistance to tension.
- **Yield Strength:** The yield strength is the pressure at which the material begins to display extremely durable distortion or pliancy. It shows the place where the material changes from a flexible to a plastic way of behaving.

- **Modulus of Versatility:** The material's stiffness, or its capacity to resist deformation when stressed, is measured by its modulus of elasticity, also known as Young's modulus. The linear elastic region depicts the slope of the stress-strain curve.

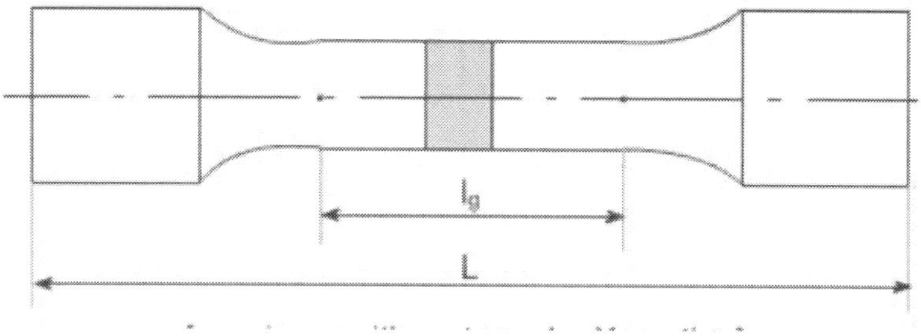

Figure 5: A rectangular X-section

Pliable testing gives urgent data for material portrayal, quality control, and plan contemplations. It aids scientists and engineers in comprehending the mechanical behavior of materials, determining whether or not they are suitable for particular applications, and ensuring compliance with industry standards. Engineers can make well-informed decisions regarding the selection of materials, structural design, and performance optimization by analyzing the stress-strain curve and determining the mechanical properties, such as ultimate tensile strength, yield strength, and modulus of elasticity.

3.2 Compression Testing

The mechanical testing technique known as compression testing is used to assess the properties and behavior of materials when subjected to compressive loads. It is a fundamental test used in engineering, manufacturing, construction, materials science, and other fields. When a material is put through compression testing, a lot can be learned about its strength, stiffness, and resistance to deformation.

The application of a compressive load to a test specimen and the evaluation of its response is the primary goals of compression testing. Compressive strength, elastic modulus, yield strength, and deformation characteristics can all be determined with this assistance.

The test is ordinarily conducted utilizing an all-inclusive testing machine (UTM) furnished with specific pressure plates that safely hold the test specimen. Depending on the particular requirements and standards, the specimen is frequently prepared in the shape of a cube, cylinder, or prism.

During the pressure test, the example is set between the pressure plates of the UTM, guaranteeing appropriate arrangement along the stacking pivot. The specimen is compressed by the machine, which gradually increases the load until a predetermined point. Throughout the test, the UTM keeps track of the applied load as well as the corresponding deformation or displacement.

- **Compressive Strength:** The maximum load or stress that a material can withstand before breaking under compression is known as its compressive strength. It is a crucial parameter for determining the material's resistance to fracture under compressive loads.
- **Flexible Modulus:** The versatile modulus, otherwise called youthful modulus in pressure, is a proportion of the material's firmness or its protection from disfigurement under applied compressive pressure inside the flexible reach. It reflects the material's capacity to return to its original shape after compression and the ratio of stress to strain.
- **Yield Strength:** The stress at which the material begins to exhibit plastic deformation or permanent strain is known as the yield strength in compression. It indicates when the material's behavior changes from elastic to plastic under compression.
- **Characteristics of Deformation:** The material's deformation under compressive loads can be better understood through compression testing. It aids in determining the onset of buckling, compression creep, or other deformations that have the potential to impact the material's performance.

Materials used in applications where compressive loads are prevalent, such as in structural components, columns, concrete, and packaging materials, necessitate compression testing. The experimental outcomes empower architects and researchers to evaluate the material's compressive strength, solidity, and reasonableness for explicit applications. It ensures that components can withstand the anticipated compressive forces and maintain their structural integrity by assisting in the selection of materials, quality control, and structural design.

By and large, pressure testing plays a pivotal role in figuring out the mechanical behavior of materials under compressive loads. It gives significant information for material portrayal, quality confirmation, and execution assessment, assisting engineers with pursuing informed choices and enhancing plans for pressure-sensitive applications.

Flexural Testing

Flexural testing, otherwise called bowing testing, is a mechanical testing strategy used to assess the behavior and properties of materials under twisting loads. In bending applications, it is frequently utilized to evaluate the strength, stiffness, and deformation characteristics of materials. Materials used in plates, beams, and other structural components that are subjected to bending forces call for special attention during flexural testing.

A test specimen is supported at two points, or simply supported on one end and subjected to a load on the other, during the flexural test. The specimen bends as the load is gradually applied until it fractures or reaches a predetermined deflection. The avoidance and applied load are estimated through the test.

Flexural testing gives a few significant mechanical properties:

- **Flexural Strength:** The maximum stress that a material experience at its outermost fibers before it fractures under bending is its flexural strength, which is also known as its modulus of rupture. It shows that the material can withstand bending forces without breaking.
- **Flexural Modulus:** The flexural modulus is a proportion of the material's solidness in bowing. Within the material's elastic range, the slope of the stress-strain curve determines it. The material's flexural modulus tells how much it will deform under a certain bending load.
- **Strain in the Flexors:** Flexural strain alludes to the deformity or lengthening of the material in light of bowing. It gives data about the material's capacity to endure avoidance and recuperate its shape in the wake of bowing.

The characterization of materials like metals, polymers, composites, and ceramics all greatly benefits from flexural testing. It aids in structural design, material selection, and quality control by assessing the material's behavior under realistic loading conditions. In bending applications, engineers and researchers rely on flexural testing to determine the strength, stiffness, and dependability of materials.

Shear Testing

The mechanical testing technique known as shear testing is used to assess the behavior and strength of materials when subjected to shear loads. When the material is subjected to forces that cause one layer or part of the material to slide or deform concerning another layer, the material's resistance to deformation is measured. Understanding the structural integrity and performance of materials in shear-sensitive applications necessitates shear testing.

A test specimen is subjected to shear forces parallel to its cross-sectional plane during shear testing. Depending on the requirements, the specimen could be a shear coupon, bonded joint, or shear test fixture. The amount of force applied increases gradually until the material fails or exhibits significant shear deformation.

Shear testing gives significant mechanical properties:

- **Strength at Shear:** Shear strength is the most extreme pressure or power that a material can endure before it flops under shear stacking. It addresses the material's capacity to resist deformity or break when exposed to shearing powers.
- **Shear Modulus:** The shear modulus, otherwise called the modulus of inflexibility, is a proportion of the material's protection from shear disfigurement. It relates the shear pressure to the shear strain and describes the material's solidity under shear stacking.

Aerospace, automotive, manufacturing, and civil engineering are just a few of the many sectors in which shear testing is frequently utilized. It aids in the design and evaluation of materials for shear-force-related applications like fasteners, adhesives, welds, and structural connections. Engineers can gain a better understanding of a material's response to shear stress, optimize designs for components that are sensitive to shear, and guarantee the integrity and safety of structures with the help of shear testing.

Hardness Testing

Hardness testing is a type of mechanical testing that is used to measure a material's resistance to indentation, scratching, or penetration. It provides useful data about the mechanical properties of a material, such as its toughness, wear resistance, strength, and overall durability. Hardness testing is used a lot in engineering, manufacturing, metallurgy, quality control, and manufacturing.

There are different hardness testing strategies accessible, each reasonable for various kinds of materials and applications. Some commonly utilized hardness testing strategies include:

- **Test of Rockwell Hardness:** Under a particular load, the Rockwell hardness test measures an indenter's depth of penetration. Depending on the material being tested, it makes use of various scales and indenters. The Rockwell hardness is not set in stone by the distinction from top to bottom when applying the test load.
- **Test of Brinell Hardness:** When performing the Brinell hardness test, a spherical indenter is subjected to a predetermined load, and the diameter of the resulting indentation is measured. The surface area of the indentation and the applied load are used to calculate the Brinell hardness value.

- **Vickers Hardness Test:** The Vickers hardness test utilizes a pyramidal-molded precious stone indenter to make a space. The Vickers hardness value is determined based on the applied load and the indentation's surface area, and the diagonal lengths of the indentation are measured.
- **Knoop Hardness Test:** Similar to the Vickers test, the Knoop hardness test makes use of a diamond indenter with a rhombic shape. It is frequently utilized for estimating the hardness of fragile materials or delicate coatings.
- **Shore Hardness Test:** The Shore hardness test is usually utilized for determining the hardness of elastomers, rubbers, and plastics. Using a durometer, the resistance to indentation is measured, and the result is a numerical hardness value.

The ability of a material to withstand localized stress, resist deformation, and perform under a variety of loading conditions can be learned from hardness testing. It aids in process optimization, quality control, and material selection. Hardness testing is used by engineers and researchers to check that material is suitable for particular applications, that heat treatments or surface treatments work well, and that industry standards are followed.

Impact Testing

A mechanical testing technique called impact testing is used to determine a material's capacity to resist fracture and absorb energy under high-velocity loading conditions. It is especially significant for materials that are inclined to unexpected disappointment or exposed to effects and dynamic stacking, like metals, composites, plastics, and ceramics.

A pendulum, drop weight, or other impact device makes a high-velocity impact on a test specimen during impact testing. The effect energy is estimated, and the way the material behaves during influence is noticed. The material's toughness, ductility, resistance to brittle fracture, and capacity to absorb energy before fracture are all revealed by the test.

Ordinarily, utilized influence testing techniques include:

- **Charpy Effect Test:** A pendulum is used to hit a notched or V-notched object in the Charpy impact test. How much energy was consumed by the example before cracking is estimated. This test is frequently used to determine a material's impact toughness.
- **Impact Test of Izod:** The specimen configuration used in the Izod impact test is different from that of the Charpy test. It measures the amount of energy the specimen takes in when it is struck.

The behavior of a material under sudden or dynamic loading conditions can be studied in detail through impact testing. It assists specialists and scientists with assessing the material's protection from breakage, its capacity to ingest influence energy, and its reasonableness for applications where effect loads are predominant, like underlying parts, auto parts, apparatus, and wellness gear.

Fatigue Testing

Weakness testing is a mechanical testing technique used to assess the solidity and execution of materials under cyclic stacking conditions. It re-creates the cycle of loading and unloading that materials go through over their lifetime. Weariness testing is vital for surveying a material's protection from weakness disappointment and deciding its exhaustion strength, weariness life, and weakness break development qualities.

A test specimen is subjected to cyclic loading with varying stress amplitudes in fatigue testing. Depending on the particular requirements of the application, the loading can be applied in tension, compression, or bending. Until the specimen fails or reaches a predetermined number of cycles, the cyclic loading continues.

The weariness test results give a few significant boundaries:

- **Strength of fatigue:** The maximum stress amplitude that a material can withstand for an infinite number of cycles without fatigue failure is known as fatigue strength or endurance limit. It addresses the material's capacity to resist exhaustion under cyclic stacking.
- **Life in Fatigue:** The number of cycles a material can withstand at a given stress amplitude before a fatigue failure is reached is known as fatigue life. It shows how long the material will last when subjected to cyclic loading.
- **Growth of fatigue cracks:** Additionally, fatigue testing aids in comprehending the development of fatigue cracks within the material. It reveals the material's capacity to withstand crack growth under cyclic loading as well as the rate of crack propagation and initiation.

Weariness testing is fundamental for materials utilized in businesses like aviation, auto, development, and hardware, where parts are exposed to continued stacking. It permits specialists to evaluate the weaknesses of materials, streamline plans to upgrade weariness resistance, and guarantee the drawn-out dependability and wellbeing of designs.

Creep Testing

A mechanical testing technique called creep testing is used to assess a material's behavior when subjected to constant or varying loads for an extended period at high temperatures. Creep is the gradual, time-dependent deformation of a material when it is put under constant stress for an extended period. Understanding the material's deformation characteristics, such as its creep rate, creep resistance, and time to failure under particular loading and temperature conditions, is made easier by creep testing.

In creep testing, a test specimen is subjected to constant stress or load for an extended period at an elevated temperature. Over time, the specimen's deformation or strain is continuously tracked. The material's creep behavior, including the primary, secondary, and tertiary creep stages, can be observed during the test.

EXPERIMENTAL APPARATUS AND PROCEDURES

Instrumentation for Mechanical Testing

Mechanical testing relies heavily on instrumentation because it enables precise measurements of various parameters and guarantees reliable and consistent results. Depending on the particular requirements of the testing, various kinds of instruments and equipment are utilized. Some common instruments for mechanical testing are as follows:

- **Charge Cells:** During mechanical testing, the applied force or load is measured with load cells. They convert the mechanical power into an electrical signal, which can be recorded and dissected. Load cells come in different plans, including strain check load cells and piezoelectric burden cells, and are chosen because of their ideal reach and exactness of power estimation.
- **Extensometers:** During mechanical testing, extensometers are used to measure changes in specimen length or strain. They enable the calculation of mechanical properties like Young's modulus, yield strength, and ultimate tensile strength because they provide precise strain measurements. Extensometers can be appended to the test example straightforwardly or in a roundabout way through an outer edge.
- **Removal Sensors:** Relocation sensors, Linear Variable Differential Transformers (LVDTs), or laser uprooting sensors, are utilized to quantify the dislodging or deformity of the test example during testing. They inform about the material's behavior in deformation, such as elongation, compression, or bending.

- **Strain Metres:** Strain gauges are used to directly measure strain on the test specimen's surface. They consist of a fine wire or foil framework that twists with the material, and the subsequent strain changes the electrical resistance of the measure. Strain measures are generally utilized for pressure strain examination, weariness testing, and material portrayal.
- **Transducers of Pressure:** During testing, pressure or stress applied to a material is measured with pressure transducers. The pressure is transformed into an electrical signal that can be recorded and examined by them. Compression testing, impact testing, and hardness testing all make frequent use of pressure transducers.
- **Sensors for Temperature:** Temperature sensors, like thermocouples, are utilized to gauge the temperature of the test climate or the example during testing. Because it affects the mechanical behavior of materials tested at elevated or extreme temperatures, temperature monitoring is essential.
- **Systems for Data Acquisition:** During mechanical testing, data from various sensors is captured, stored, and analyzed by data acquisition systems. These frameworks comprise equipment and programming parts that consider continuous information recording, perception, and examination. They make it possible for scientists and engineers to keep an eye on test parameters, collect data, and derive useful insights from test results.
- **Control Frameworks:** Control systems are utilized to regulate the applied load, displacement, or strain rate in some mechanical testing setups. The testing conditions can be precise and repeatable thanks to these systems, which guarantee precise control over the parameters.
- **Systems for Imaging and Visualisation:** Imaging and representation frameworks, like fast cameras, magnifying instruments, or filtering electron magnifying instruments (SEM), are utilized to notice and break down the way the test example behaves during testing. Visual information on material deformation, fracture surfaces, and microstructural changes is provided by these systems.
- **Tools for Analysis and Software:** In mechanical testing, data processing, analysis, and reporting necessitate the use of software and analysis tools. They make it easier to estimate fatigue life, generate stress-strain curves, calculate mechanical properties, and statistically analyzed test results.

To guarantee accurate and reliable measurements, mechanical testing instrumentation must be calibrated and checked for accuracy. When conducting mechanical testing experiments, it is essential to select, calibrate, and maintain the appropriate instruments.

Sample Preparation and Specimen Geometry

The choice of fitting example math and appropriate example planning strategies is fundamental to guaranteeing delegation and reliable testing conditions. Here are key contemplations for test readiness and examples of math in mechanical testing:

- **Material Determination:** The material selection is crucial and ought to be representative of the application or subject matter under study. When selecting the material, consideration should be given to its properties, intended application, and testing objectives.
- **Measurement Geometry:** The specific test that is being carried out should be used to define the specimen's geometry. Normal example calculations incorporate round and hollow, rectangular, or kaleidoscopic shapes. The specimen's dimensions and proportions, such as length, width, and thickness, ought to be in line with the relevant test standards or particular requirements for research.
- **Preparing the Surface:** The example surfaces ought to be painstakingly ready to guarantee appropriate holding, limit surface blemishes, and diminish the impact of surface harshness on test results. This might include machining, crushing, cleaning, or other surface treatment methods.
- **Example Mounting:** Contingent upon the test technique, the example might be mounted or clasped in a reasonable installation to guarantee a legitimate arrangement and secure connection. For tests involving tension, compression, bending, or shear loading, this is especially crucial.
- **Example Direction:** For some tests, the specimen's orientation concerning the material's microstructure or a particular loading direction may be crucial. To ensure meaningful and consistent results, it needs to be carefully controlled and documented.
- **Volume and Size of the Sample:** The specimen's volume and size should be chosen to represent the intended use or research. To ensure the validity of test results and minimize the impact of statistical variations, a sufficient sample size is required.
- **Example Readiness Procedures:** Depending on the material and the requirements of the test, sample preparation techniques may include cutting, machining, grinding, or other techniques. During the preparation process, care must be taken to minimize induced stresses, surface damage, or changes in the material's properties.

- **Example Moulding:** To ensure consistent behavior and eliminate any effects of previous processing or environmental exposure, some materials may require specific conditioning procedures, such as heat treatment or moisture equilibration.

Reproducibility and traceability of test results depend on accurate documentation of sample preparation, including information on the material composition, specimen dimensions, surface preparation methods, and any conditioning steps.

It is vital to observe pertinent test guidelines and conventions to guarantee consistency and likeness across various examinations. Meaningful comparisons and a better comprehension of material behavior under specific loading conditions can be achieved by adhering to standard guidelines for specimen geometry and sample preparation.

Test Standards and Protocols

In mechanical testing, results must be consistent, comparable, and reliable by adhering to standardized test methods and protocols. Test principles are created by perceived associations like ASTM Worldwide (previously known as the American Society for Testing and Materials), ISO (the Global Association for Normalisation), and other public and global guidelines bodies. Specific mechanical tests, the specification of test parameters, specimen preparation, and data analysis are all outlined in these standards. Some common test protocols and standards are as follows:

- **Tensile Testing of Metallic Materials:** ASTM E8/E8M The procedures for carrying out tensile tests on metallic materials are outlined in this standard. It defines the specimen's geometry, how to test it, and how to calculate mechanical properties like elongation, ultimate tensile strength, and yield strength.
- **ASTM D638:** Elastic Testing of Plastics: The test procedure for determining plastics' tensile properties is outlined in this standard. The specimen's geometry, the procedure for testing, and the calculation of mechanical properties like tensile strength, elongation, and modulus of elasticity are all outlined in this document.
- **Flexural Properties of Reinforced and Unreinforced Plastics:** ASTM D790 The test procedure for determining the flexural properties of composite materials and plastics is outlined in this standard. It describes how to prepare specimens, conduct tests, and calculate flexural modulus, flexural strength, and flexural strain.
- **Micro indentation Hardness Testing:** ASTM E384 This standard indicates the test strategy for deciding the micro indentation

hardness of materials utilizing different hardness testing methods, like Vickers and Knoop. It explains how to prepare the specimen, conduct the test, and calculate the hardness values.

- **Axial Fatigue Testing of Metallic Materials:** ASTM E466 Axial fatigue testing of metallic materials is outlined in this standard. The purpose of determining the properties of fatigue strength, fatigue life, and fatigue crack growth, includes specimen preparation, a test procedure, and data analysis.
- **Tensile Testing of Metallic Materials:** ISO 6892-1 This worldwide standard defines the test technique for the malleable testing of metallic materials. Similar to ASTM E8/E8M, it covers specimen preparation, test procedure, and mechanical property calculation. ISO 527: Tensile Testing of Plastics This worldwide standard indicates the test technique for determining the malleable properties of plastics. It gives instructions on how to prepare a specimen, conduct a test, and figure out the tensile modulus, tensile strength, and elongation.

Researchers and engineers ensure that their results can be compared to data from other studies or industries by adhering to standardized test protocols. This makes it easier to comprehend and evaluate the performance of materials. Test guidelines likewise aid quality control, material affirmation, and consistency with industry guidelines.

To ensure that the specimen geometry, test setup, instrumentation, and data analysis procedures are correct, it is essential to consult the relevant test standards and protocols for particular mechanical tests. Mechanical testing is guaranteed to be repeatable and accurate as a result of this, which aids in obtaining consistent and accurate results.

Data Acquisition and Analysis

The cycle includes catching and recording information during the test, performing information examination utilizing suitable procedures, and deciphering the outcomes. Here are the vital parts of information procurement and examination in mechanical testing:

- **Method of Data Acquisition:** During mechanical testing, data must be captured and recorded using a dependable data acquisition system. It typically includes hardware components for data storage, such as data loggers or computer interfaces and sensors, transducers, amplifiers, and so on. The framework ought to be fit for catching pertinent boundaries like power, uprooting, strain, temperature, and time with proper examining rates to guarantee a precise portrayal of the test conditions.

- **Calibration:** To guarantee accurate and dependable measurements, each sensor and measuring device should be properly calibrated before the test. Adjustment includes contrasting the result of the estimating instrument against a known reference or standard. This step is vital for dispensing with estimation mistakes and guaranteeing the recognizability of the information.
- **Information Examination Procedures:** The collected data are analyzed using a variety of methods to find relevant information. These strategies might incorporate factual investigation, signal handling, bend fitting, Fourier examination, and other numerical techniques. The decision of examination strategies relies upon the particular goals of the test and the idea of the information being dissected.
- **Analysis of the Statistic:** The summarization and interpretation of the data, the identification of trends, and the evaluation of the results' reliability and significance all benefit from statistical analysis. Mean values, standard deviations, confidence intervals, hypothesis testing, and regression analysis might all be part of it.
- **Fitting curves:** Bend fitting procedures are utilized to fit numerical models to exploratory information, considering the assurance of material properties, like yield strength, modulus of flexibility, or weariness life. Relapse investigation and streamlining calculations are frequently utilized to decide the best-fit boundaries.
- **Visualization:** The data is presented clearly and concisely through the use of visualization strategies like graphs, charts, and plots. They make it easier to interpret the results by pointing out patterns, trends, and anomalies in the data.
- **Conclusions and Interpretations:** Because of the information examination, scientists can decipher the outcomes and make determinations concerning the material's mechanical way of behaving, execution, and properties. This includes comparing the results to relevant standards or specifications and determining mechanical properties, failure modes, and performance under various loading conditions.
- **Reporting and documentation:** It is crucial to report all parts of information procurement and examination, including test conditions, hardware utilized, information handling strategies, and translation of the outcomes. A complete report ought to be ready, featuring the key discoveries, investigation methodology, and ends drawn from the information.

Researchers and engineers can gain insight into the behavior of materials, validate design concepts, improve performance, and guarantee the quality and dependability of products by using effective data acquisition and analysis methods. It facilitates continuous improvement, enables informed decision-making, and contributes to the advancement of mechanical engineering knowledge.

Material Characterization

It includes the review and evaluation of material properties and conducts to comprehend how materials respond to different mechanical powers and natural circumstances. The suitability of materials for particular applications, their performance under various loading conditions, and the design and optimization of structures and components can all be guided by material characterization. Material portrayal includes a scope of procedures and tests that give experiences into different material properties. Common approaches to material characterization include:

- **Tests on the Mechanical:** Materials' responses to controlled forces are measured during mechanical testing. Among the most prevalent mechanical tests are tensile, compression, flexural, and hardness tests. Material strength, stiffness, ductility, and deformation resistance are all measured by these tests.
- **Microstructural Examination:** The internal structure of materials at the microscopic level is the subject of microstructural analysis. The grain structure, phase composition, and defects in materials are examined using optical microscopy, electron microscopy (scanning electron microscopy and transmission electron microscopy), and X-ray diffraction. Material homogeneity, crystallographic orientation, and the presence of microstructural features like grain boundaries, dislocations, and precipitates are all revealed through microstructural analysis.
- **Warm Investigation:** Differential scanning calorimetry (DSC) and thermogravimetric analysis (TGA) are examples of thermal analysis methods that look at how materials respond to temperature changes. Phase transitions, thermal stability, thermal expansion, and heat capacity are all covered by them.
- **Testing for Rheology:** Materials' viscoelastic and flow behavior are evaluated through rheological testing. Measurements of viscosity, shear modulus, and damping properties are made with tools like rotational rheometry and dynamic mechanical analysis (DMA). Rheological information helps in understanding the material reaction to mishappening and its capacity to endure dynamic burdens.
- **Analytical Chemistry:** To determine the chemical composition, elemental composition, and molecular structure of materials,

spectroscopy (for example, Fourier-transform infrared spectroscopy and X-ray photoelectron spectroscopy) and chromatography are used in chemical analysis. Material purity, elemental impurities, and the presence of functional groups or contaminants are all revealed through chemical analysis.

- **Break Mechanics:** The study of materials' behavior under stress concentrations and their resistance to crack propagation is known as fracture mechanics. Strategies like break sturdiness testing and weakness testing assist with deciding material durability, exhaustion strength, and protection from break inception and development.
- **Testing with Tribology:** Materials' friction, wear, and lubrication characteristics are evaluated through tribological testing. Material performance under sliding, rolling, and abrasive conditions are evaluated using methods like wear testing and tribometer testing. Tribological data aids in the selection of the best materials and the creation of efficient lubrication systems.
- **Testing in the Environment:** Material durability and degradation behavior are evaluated through environmental testing by subjecting them to various environmental conditions like temperature, humidity, and chemical exposure. To simulate and evaluate material performance under actual conditions, environmental chambers, accelerated aging tests, and corrosion testing is utilized.

By joining various portrayal procedures, specialists can get a thorough comprehension of material properties, conduct, and execution. The information acquired from material portrayal is utilized to approve hypothetical models, adjust reproductions, and guide material determination and plan choices. In fields like aerospace, automotive, biomedical, and structural engineering, where the performance and dependability of materials are crucial to the application's success, material characterization is essential.

ADVANCED TECHNIQUES AND RECENT DEVELOPMENTS

Nanoindentation and Microscale Testing

Advanced methods like nanoindentation and microscale testing are used to study the mechanical properties of materials at very small length scales. Due to their capacity to provide useful insights into the behavior of materials at the micro and nanoscales, these methods have received a lot of attention in materials science and engineering. Nanoindentation involves measuring the depth of penetration achieved by applying a controlled force to a small indenter probe, typically a diamond tip. Microscale testing alludes to a scope of mechanical tests performed on limited scope examples, for example, micro tensile testing or micro compression testing.

In comparison to conventional macroscale testing techniques, nanoindentation and microscale testing offer several advantages. First, they make it possible to take precise, local measurements of material properties in particular phases or regions. This is especially important in heterogeneous materials, where the properties might shift across various microstructural highlights. Second, these methods make it possible to characterize small-scale structures, coatings, and thin films that are difficult to test using standard methods.

The responses of materials at the micro and nanoscale, such as hardness, elastic modulus, plasticity, and creep behavior, can be learned from the results of nanoindentation and microscale testing. These properties assume a significant role in grasping material exhibition, like wear obstruction, bond, and weakness strength. Furthermore, nanoindentation considers the assurance of nanomechanical properties, for example, the hardness-to-modulus proportion, which gives insight into material misshapen Ing instruments and can be utilized to survey material conduct under outrageous circumstances.

Often, sophisticated mathematical models and analysis methods are used to interpret the data from nanoindentation and microscale testing. Hardness, elastic modulus, and plasticity indices are typically derived from indentation load-displacement curves through analysis. Different models, like the Oliver-Pharr strategy or the flexible plastic limited component investigation, are utilized to precisely determine these properties.

Nanoindentation and microscale testing track down applications in a large number of fields, including materials science, surface design, microelectronics, and biomaterials. They are utilized to explore the mechanical behavior of dainty movies, coatings, microelectromechanical frameworks, organic tissues, and other microscale structures. Understanding and optimizing the mechanical properties of materials in nanotechnology, where control over material behavior at small scales is crucial, can greatly benefit from these methods.

They give important bits of knowledge into the way materials behave in limited-scope frameworks, empowering scientists and specialists to plan and upgrade materials for explicit applications. As progressions keep on being made in instrument planning and information examination strategies, nanoindentation and microscale testing will additionally contribute to the comprehension and improvement of cutting-edge materials with custom-made properties at small length scales.

Digital Image Correlation (DIC)

In solid mechanics and materials engineering, it is frequently used to investigate the behavior of structures, components, and materials under a variety of loading conditions. To precisely measure strain and deformation, DIC makes use of high-resolution cameras and image processing algorithms to follow the movement of markers or patterns on the material surface.

During the loading process, a series of images of the material surface are taken, which is the fundamental principle of DIC. Most of the time, these pictures are taken with a high-quality lens on a digital camera. A predefined grid or a random speckle pattern is applied to the specimen's surface before testing. The spot design fills in as the reference for following the surface uprooting and twisting.

The essential steps of the DIC procedure are as follows:

- **Acquisition of Images:** At various stages of loading, the digital camera takes high-resolution images of the specimen surface. The pictures ought to have adequate picture quality and aim to precisely follow the spot example or framework.
- **Pre-processing of images:** Pre-processing is used to improve image quality, reduce noise, and increase contrast in the acquired images. Filtering, thresholding, and image enhancement algorithms might be used in this step.
- **Correlation of Images:** Correlation algorithms are used to compare the pre-processed images to determine the speckle pattern's or grid's displacement and deformation. The calculations look for examples and match them between the reference picture and the resulting pictures, ascertaining the relocation field.
- **Strain Estimation:** Utilizing the appropriate mathematical formulations, strain values are calculated from the measured displacement field. It is possible to determine strain components like normal strains (axial and transverse) and shear strains, making it possible to characterize intricate deformation patterns.
- **Information Examination and Representation:** The strain and relocation information obtained from DIC is dissected and imagined utilizing particular programming. Strain shapes, relocation plots, and distortion guides can be produced to envision the material's reaction under different stacking conditions.

Material characterization and mechanical testing benefit greatly from DIC. It is a non-contact method that minimizes specimen disturbance and eliminates the requirement for physical strain gauges or extensometers. Moreover, DIC gives full-field estimations, empowering the distinguishing proof of confined strain fixations, discontinuities, and material deformities. It can be used on a wide range of materials, including biological tissues, polymers, composites, and metals.

In solid mechanics research and engineering, DIC has many uses. It is utilized in component validation, fatigue analysis, impact testing, structural testing, material characterization, and fatigue analysis. DIC has demonstrated that it is especially helpful in concentrating on the mechanical behavior of mind-boggling structures, for example, aviation parts, car body boards, and biomedical inserts. It helps to validate numerical simulations, improve designs, and comprehend the mechanisms of material failure.

In-situ Mechanical Testing

Under real-time or simulated operating conditions, material behavior can be directly observed and characterized using this method. In-situ testing gives significant bits of knowledge into how materials and designs respond to outside upgrades and empowers the investigation of peculiarities like misshapen Ing, breakage, and exhaustion in a controlled climate.

Depending on the study's specific requirements, in-situ mechanical testing can be carried out using a variety of methods. Methods for in-situ testing include:

- **Tensile Testing in the Field:** Elastic testing is performed while at the same time noticing the material's reaction using procedures like optical microscopy, examining electron microscopy (SEM), or X-beam imaging. The investigation of phenomena like necking, strain localization, and the initiation and propagation of fractures is made possible by this.
- **Compression Testing in the Field:** Imaging methods and compression tests can be used to keep an eye on changes in sample morphology, deformation mechanisms, and failure modes. This is especially useful for studying how foams, cellular structures, and porous materials behave.
- **Fatigue Testing in the Field:** Exhaustion tests performed under in-situ conditions give insights into the commencement and development of weakness breaks, break conclusion peculiarities, and the impact of cyclic stacking on material properties. This can be accomplished by utilizing specific stacking apparatuses or gear that takes into consideration the use of cyclic burdens while checking break proliferation utilizing procedures like acoustic emanation or advanced picture relationship.

In-situ mechanical testing offers a few benefits over customary ex-situ testing strategies. By simulating the actual conditions under which it operates, it provides a comprehension of material behavior that is more realistic. In-situ testing can uncover experiences that might be missed in regular testing, like the impact of temperature, mugginess, or time-subordinate peculiarities on material execution. In addition, in-situ testing enables a direct correlation between applied loads and observed responses, facilitating a deeper comprehension of the mechanical mechanisms at work.

Computational Modeling and Finite Element Analysis

Computational demonstration and Finite Element Analysis (FEA) are imperative devices in strong mechanics exploration and design. The numerical method known as FEA is used to simulate the behavior of structures and materials under a variety of loading conditions. To ascertain the material's stress, strain, and deformation distributions, involves discretizing the structure into finite elements and solving the governing equations of motion.

In the study of solid mechanics, computational modeling and FEA offer several advantages:

- **Optimization of Design:** FEA can be utilized to upgrade the plan of designs or parts by assessing their presentation under various stacking situations. It makes it possible to investigate a variety of design configurations, material choices, and boundary conditions to boost structural integrity, reduce weight, or improve performance.
- **The expectation of material's behavior:** Predicting the behavior of a material under intricate loading conditions, which may be challenging or costly to test experimentally, is made possible by FEA. It is capable of simulating nonlinear behavior, plasticity, viscoelasticity, and other properties of materials, giving useful insights into how materials react to real-world conditions.
- **Disappointment Examination:** By predicting failure modes, locating critical stress concentrations, and identifying the factors that contribute to failure, FEA can assist in failure analysis. It empowers architects to figure out disappointment systems, further develop plans, and improve the unwavering quality of designs or parts.
- **Studies in Parametric:** Parametric studies can be carried out using FEA, in which input parameters like geometry, material properties, or boundary conditions are changed to run multiple simulations. Engineers can use this information to determine the sensitivity of the structure's response to various factors and adjust their design decisions accordingly.

However, it is essential to keep in mind that both FEA and computational modeling are constrained by the need for precise material models and boundary conditions for accurate predictions.

CHALLENGES AND FUTURE DIRECTIONS

Integration of Experimental and Computational Approaches

In non-disastrous assessment, there is a developing pattern towards coordinating trial and computational ways to upgrade the review cycle and further develop exactness. Computational methods, such as finite element analysis and computational fluid dynamics, can simulate the behavior of materials and structures, whereas experimental techniques provide data and measurements in real-time. By consolidating trial and computational procedures, it is feasible to get a more exhaustive comprehension of the examined item's way of behaving, foresee disappointment modes, improve investigation boundaries, and work with dynamic cycles.

Before carrying out actual inspections, virtual testing and simulations can be conducted as a result of this integration, saving money and time. It likewise empowers the improvement of cutting-edge calculations and AI procedures to robotize deformity identification and investigation, working on the effectiveness and dependability of non-damaging assessment.

Standardization and Certification

Non-destructive evaluation relies heavily on standardization and certification to guarantee reliable and consistent inspection procedures across industries. Different associations, like the American Society for Non-destructive Testing (ASNT) and the International Organization for Standardization (ISO), create and distribute principles and rules for NDE procedures.

Procedures, acceptance criteria, equipment calibration requirements, personnel qualifications, and reporting formats are all defined by these standards. They contribute to the assurance that inspections are carried out appropriately, with the right tools and skilled workers, and that the outcomes are consistent, comparable, and traceable.

Confirmation programs are additionally accessible to survey and check the capabilities of people performing non-disastrous assessments. To guarantee that employees have the knowledge and abilities necessary to carry out inspections per the required standards, these programs typically include examinations, periodic recertification, and training.

Standardization and certification promote safety, quality assurance, and adherence to industry best practices by giving industries, regulators, and customers assurance that non-destructive evaluations are carried out dependably and consistently.

CONCLUSION

In solid mechanics, the study of elasticity and mechanical testing shed light on the behavior and properties of materials when subjected to external forces. We have gained a deeper understanding of how materials deform and respond to mechanical loading by analyzing stress-strain relationships. The characterization of materials' reversible deformation has relied heavily on the idea of elasticity. We can quantify the material's capacity to withstand stress and return to its original shape after the load has been removed by studying the elastic modulus, which includes Young's modulus, shear modulus, and bulk modulus. Predicting how well structures will perform under a variety of loading conditions and designing them are both dependent on this knowledge.

The evaluation of materials' mechanical properties relies heavily on mechanical testing. Tractable testing, pressure testing, and shear testing have given significant information on the strength, flexibility, and sturdiness of materials. These tests have empowered specialists and analysts to survey the primary trustworthiness of parts and make informed choices concerning the material determination and plan improvement. In addition, mechanical testing procedures, for example, hardness testing and effect testing, have permitted us to evaluate material protection from the entrance and unexpected stacking separately. These tests are especially important in applications where materials need to endure grating or high-influence conditions.

Summary of Key Findings

The vital discoveries in the field of non-horrendous assessment feature the meaning of different procedures and approaches in guaranteeing the respectability, unwavering quality, and well-being of materials and designs. Visual examination fills in as a major strategy for surface imperfection evaluation, while fluid penetrant testing and attractive molecule testing are powerful in recognizing surface-breaking surrenders. Radiographic testing enables the examination of hidden structures, whereas ultrasonic testing enables the thorough inspection of internal flaws. The coordination of trial and computational methodologies has emerged as an important instrument, working with virtual testing, recreations, and prescient demonstrating. Guidelines and assurance of consistent practices are provided by standardization and certification initiatives. Versatility and mechanical testing assume a significant role in material portrayal, deformity identification, underlying well-being checking, quality control, material certification, and plan improvement. These key discoveries underscore the ceaseless progressions and the requirement for a far-reaching way to deal with the non-damaging assessment, guaranteeing the dependability, well-being, and execution of materials and designs in different enterprises.

Importance of Elasticity and Mechanical Testing

Flexibility and mechanical testing are vital in the field of non-disastrous assessment because of multiple factors:

- **Characteristics of the Material:** Testing materials for elasticity and mechanical properties yields useful information about their mechanical properties. These properties incorporate modulus of versatility, yield strength, extreme elasticity, malleability, and crack sturdiness. For evaluating the structural integrity, dependability, and performance of materials and components, it is essential to comprehend these properties.
- **Sizing and Detection of Defects:** Defects in materials like cracks, voids, delamination, and inclusions can be identified and characterized using elasticity and mechanical testing methods. By exposing the material to controlled pressure or strain, variations in the mechanical reaction can show the presence and seriousness of imperfections. With this information, it is possible to take the appropriate actions for maintenance or repair by determining the size, location, and criticality of the defects.
- **Monitoring the Health of the Structure:** In structural health monitoring, elasticity, and mechanical testing methods are utilized to evaluate the condition and behavior of structures over time. It is possible to detect and track the progression of damage, degradation, or fatigue in structures by periodically measuring changes in mechanical properties like stiffness or damping. This empowers proactive support, fixes, or substitutions, limiting the risk of devastating disappointments.
- **Quality Control and Material Confirmation:** Flexibility and mechanical testing play an urgent role in quality control processes during material creation and assembly. These tests guarantee that materials meet specified standards and requirements for mechanical properties. They also provide crucial information for material certification, which verifies whether or not materials are suitable for particular applications and how well they perform.
- **Optimizing and designing:** Structures and components are designed and optimized with the help of elasticity and mechanical testing data. Engineers can make educated choices regarding material selection, geometry, and load-bearing capabilities by comprehending the mechanical behavior of materials. As a result, designs that meet performance requirements while minimizing weight, cost, and potential failure risks become more effective and dependable.

Because they provide crucial information about material properties, defect detection and sizing, structural health monitoring, quality control, material certification, design optimization, elasticity, and mechanical testing are essential components of non-destructive evaluation. These testing techniques contribute to guaranteeing the security, dependability, and execution of materials and designs across different enterprises.

Future Perspectives and Research Directions

- **Techniques for High-Tech Imaging:** NDE's capabilities can be enhanced by continuing research and development of cutting-edge imaging methods. To enhance defect detection and characterization in a variety of materials, this includes investigating brand-new imaging modalities like terahertz imaging, microwave imaging, and advanced tomographic techniques.
- **Multi-Scale and Multi-Modal Strategies:** A more complete evaluation of materials and structures can be achieved by combining multiple NDE techniques at various scales, like ultrasonic, electromagnetic, and optical methods. This multi-modular methodology empowers the identification and portrayal of deformities and harm instruments at different length scales, improving the exactness and dependability of NDE.
- **Automation and robotics:** In particular in complex and hazardous environments, the incorporation of robotics and automation into NDE can significantly enhance inspection accessibility and efficiency. Examination ought to zero in on creating independent or semi-independent review frameworks, including robots, crawlers, or mechanical arms, equipped for performing assessments with negligible human mediation.
- **Non-Horrendous Assessment of Cutting-edge Materials:** Advanced materials like composites, ceramics, and additive manufacturing require specialized NDE methods that have been developed specifically for these materials. Innovative NDE methods that effectively identify and characterize defects and damage in these challenging materials ought to be the primary focus of research.
- **Data Analysis and Quantitative NDE:** The quantitative analysis of NDE data can be improved by advances in data analytics, machine learning, and artificial intelligence. To speed up and improve the accuracy of NDE inspections, robust algorithms, and models that can automate defect detection, classification, and quantification should be the primary focus of research.
- **Real-time and in-service monitoring:** Continuous evaluation and early detection of material degradation or structural changes can be made possible by the development of in-service and real-time monitoring techniques. During their operational lifetimes, this can

assist in the implementation of proactive maintenance strategies, optimize asset management, and enhance the overall safety and dependability of structures.

- **Natural and maintainable Contemplations:** The environmental impact and sustainability of NDE techniques should be the focus of future research. This includes the implementation of life cycle assessment principles into NDE practices, the reduction of waste produced during inspections, and the development of eco-friendly inspection methods.

In general, the goals of non-destructive evaluation's future perspectives and research directions are to expand its applicability to advanced materials, enable real-time monitoring, improve automation and efficiency, improve inspection capabilities, and address environmental concerns. These advancements will additionally fortify the job of NDE in guaranteeing the uprightness, security, and supportability of materials and designs in different businesses.

6

Plasticity Theory and Temperature Dependence of Flow Stress of a Solid Mechanics

INTRODUCTION

A fundamental idea in solid mechanics, plasticity theory describes how materials behave when subjected to plastic deformation. Plastic deformation is the persistent change in a material's shape or size when forces are applied, typically beyond the elastic limit. Understanding the temperature dependence of stream pressure, which is the protection from plastic disfigurement, is vital for anticipating and dissecting the way materials behave in different designing applications. Several things can affect a material's flow stress, such as the temperature at which the deformation occurs. The material's atomic mobility increases at higher temperatures, enhancing its plasticity. This peculiarity is regularly seen in high-temperature cycles like metal shaping and projecting. On the other hand, the material's resistance to plastic deformation significantly improves at low temperatures, reducing its ductility and making it more brittle.

Mathematical models like the Arrhenius equation and the power-law equation are frequently utilized to quantify the relationship that exists between temperature and flow stress. These models shed light on the mechanisms governing plastic deformation and the temperature-dependent behavior of materials. Engineers and researchers can improve material selection and processing conditions to achieve the desired mechanical properties and performance by comprehending the temperature dependence of flow stress. we expect to investigate the temperature dependence of stream pressure in strong mechanics. The mechanical behavior of various materials at various temperatures will be the subject of our investigation. We will identify the underlying mechanisms governing the flow stress-temperature relationship by analyzing the experimental data and employing theoretical models. We will also talk about how temperature dependence affects engineering structure design and analysis, as well as potential applications in the manufacturing, aerospace, and automotive sectors. In general, accurately predicting material behavior, optimizing processing conditions, and ensuring the structural integrity and dependability of engineered components all require an understanding of the temperature dependence of flow stress. This study will add to the current information on strong mechanics and give important experiences for viable design applications.

Background

Many engineering materials exhibit plastic deformation as a fundamental behavior when subjected to external forces. It is described by the super-durable modification of iotas or particles inside the material, bringing about an adjustment of shape or size. The Pliancy Hypothesis gives a system for understanding and foreseeing this twisting way of behaving. The stream pressure of a material is a vital boundary in the versatility hypothesis, addressing the obstruction of the material to plastic misshapen Ing. Temperature is one factor that affects it. The temperature dependence of stream pressure is a vital perspective in strong mechanics, as it straightforwardly influences the material's mechanical properties and conduct.

At raised temperatures, the energy accessible for nuclear movement increments, permitting particles to conquer obstructions and work with plastic deformity all the more without any problem. The material's ductility and flow stress both decrease as a result of this increased atomic mobility. This phenomenon is especially important in processes like hot rolling, extrusion, and forging, in which metals are heated up to make them easier to form. On the other hand, at low temperatures, the amount of energy that can be used to move atoms decreases, which makes atoms less mobile and more resistant to plastic deformation. Under applied stresses, materials typically become more brittle and susceptible to fracture. Utilizing this temperature-dependent behavior, cold working processes like cold rolling and cold forging boost materials' strength and hardness.

Mathematical models are frequently utilized in the description of the temperature dependence of flow stress. The Arrhenius equation, which relates the temperature and activation energy to the flow stress, is one model that is frequently utilized. The flow stress is assumed to have an exponential relationship with temperature in the Arrhenius equation. The power-law equation, which describes a power-law relationship between flow stress and temperature, is another popular model. These models make it possible to estimate flow stress at various temperatures and provide a quantitative understanding of temperature dependence. Understanding the temperature dependence of stream pressure is vital in different design disciplines. It assists in planning materials and parts that, with canning, endure explicit temperature conditions, advancing assembling processes, and foreseeing the way designs behave under various warm conditions. Engineers and researchers can make well-informed choices about material selection, processing parameters, and structural design by studying the temperature dependence of flow stress, which ultimately ensures the performance and reliability of engineered systems.

Objectives

The temperature dependence of flow stress in solid mechanics and plasticity theory are studied for two reasons: predicting mechanical response under various conditions and understanding material behavior.

First and foremost, the objective is to acquire a deeper comprehension of the plasticity theory, which emphasizes the mechanical behavior of materials that are subjected to plastic deformation. The way in which materials respond to external loads, undergo irreversible deformations, and ultimately attain a new equilibrium state is the subject of plasticity theory research. The fundamental underlying mechanisms of plastic deformation, such as dislocation motion, slip, and grain boundary interactions, are the goals of research into plasticity theory. This understanding can assist in planning materials with worked-on mechanical properties, streamlining production processes, and foreseeing disappointment modes.

In solid mechanics, the second objective is to investigate the temperature dependence of flow stress. Stream pressure alludes to the opposition experienced by a material when it goes through plastic distortion. The temperature plays an essential role in determining the stream pressure of a material. Due to enhanced dislocation motion and increased atomic mobility, materials typically exhibit lower flow stresses at elevated temperatures. Interestingly, at lower temperatures, materials will generally have higher stream stresses because restricted nuclear power limits disengagement movement. Researchers hope to develop models and equations that accurately predict the mechanical response of materials at various temperatures by studying the temperature dependence of flow stress. This information is critical in different ventures, like aviation, auto, and primary design, where materials experience an extensive variety of temperature conditions during administration.

Generally speaking, the goals of concentrating on the versatility hypothesis and the temperature dependence of stream pressure in strong mechanics are to extend how we might interpret material ways of behaving and to foster prescient models that can illuminate the planning, examination, and advancement of materials and designs in different applications.

PLASTICITY THEORY

The versatility hypothesis is a major idea in strong mechanics that deals with the disfigurement of materials past their flexible breaking point. It gives a system for understanding and foreseeing the plastic deformity of materials exposed to outside loads. In this literature review, we will examine important studies and developments in plasticity theory, including its historical context, fundamental principles, mathematical formulations, and engineering applications. The early 19th century saw the beginning of research into the plastic behavior of materials, which led to the development of plasticity theory. Augustus Wilhelm von Hofmann, a German physicist, proposed the term "plasticity" in 1865 to describe the capacity of materials to undergo permanent deformation. He first proposed the idea of plasticity. In the late 19th and early 20th centuries, researchers like Tressa, Rankine, and Saint-Ven ant made significant contributions by developing theories and experiments to comprehend the plastic behavior of materials.

The behavior of materials under plastic deformation is governed by several fundamental principles, which form the foundation of plasticity theory. According to the theory of plastic deformation, when a material reaches its elastic limit, it undergoes permanent deformation without experiencing additional stress. The standard of protection of mass and similarity guarantees that the complete deformity is steady and that there are no holes or covers inside the material. In 1920, Prandtl proposed the flow rule, which states that the material's flow properties determine the plastic strain rate and that it is inversely proportional to the deviatoric stress. Constitutive equations that relate the stress and strain of a material under plastic deformation are used in the mathematical formulation of plasticity theory. The stress state at which plastic deformation begins is defined by the yield criterion, which is an important part of these equations. The Tressa criterion, the von Mises criterion, and the Mohr-Coulomb criterion, among others, have been proposed as yield criteria. These yield models give rules for determining the beginning of plastic deformity because of the pressure state and material properties.

Plasticity theory includes strain hardening, which describes the rise in material strength and resistance to deformation as plastic strain builds up, in addition, to yield criteria. The material's changing behavior during plastic deformation is captured by strain hardening models like isotropic hardening, kinematic hardening, and combined hardening. These models take into account how the material's current stress-strain response is affected by previous deformations. Structural engineering, geotechnical engineering, material science, and metal forming processes are just a few of the engineering fields where plasticity theory can be applied. In underlying design, the versatility hypothesis is utilized to examine and configure structures exposed to huge distortions, for example, plastic pivots in steel outlines or bendable conduct in tremor-safe designs. The pliancy hypothesis is additionally significant in geotechnical design for examining the solidness of soil slants, planning to hold walls, and anticipating the twisting of soil under load. Predicting material failure, studying the effects of processing and microstructure on material properties, and choosing the best materials for particular applications are all made easier with its assistance. In addition, metal forming procedures like forging, rolling, and extrusion rely heavily on plasticity theory to shape and deform materials into desired geometries by exploiting their plastic behavior.

Overview of Plasticity Theory

Plasticity theory is a subfield of solid mechanics that studies how materials behave mechanically when subjected to plastic deformation. It examines how materials respond to external loads, how they eventually reach a new equilibrium state, and whether or not permanent deformations occur. The understanding of the mechanisms by which materials deform plastically is the primary goal of plasticity theory. Plastic distortion alludes to the super-durable change in shape or size of a material that happens when the applied pressure surpasses a specific limit, known as yield strength. Not at all like flexible twisting, which is reversible and brief, plastic deformity includes the irreversible movement of separations inside the material's gem structure.

The idea of stress and strain is at the center of plasticity theory. A material's internal force per unit area is called stress, and the deformation that results from it is called strain. The pliancy hypothesis expects to lay out constitutive connections among anxiety that portray the material's mechanical reaction under various stacking conditions. Typically, these relationships are derived through mathematical modeling and experimental observations. The motion of dislocations is one of the primary phenomena that plasticity theory investigates. Separations are line deserts inside a gem cross-section that can move and collaborate, prompting plastic twisting. Versatility speculations consolidate separation movement systems, like slip and twinning, to make sense of material ways of behaving and foresee the beginning of plastic twisting.

The presence of defects or impurities, temperature, strain rate, and other factors that influence material behavior are all taken into account by plasticity theory. The material's response can be significantly affected by the strain rate, or rate of deformation. Temperature also has a significant impact on plasticity because it affects the material's ability to undergo plastic deformation and the mobility of dislocations. To describe and predict material behavior, various mathematical and computational methods are included in the field of plasticity theory. Plastic potential functions, flow rules, yield criteria, and numerical methods like finite element analysis are all examples of these. These devices permit specialists and scientists to break down and configure designs and parts with a careful comprehension of the material's plastic way of behaving. the goal of plasticity theory is to comprehend and anticipate the mechanical response of materials that undergo plastic deformation. Researchers can develop models and methods for designing materials with desired properties, optimizing manufacturing processes, and ensuring the structural integrity and safety of engineered systems by studying the mechanisms of plasticity.

Constitutive Equations

Mathematical expressions known as constitutive equations, material constitutive equations, or stress-strain relations describe the connection between a material's mechanical behavior and the external forces or deformation it experiences. These equations are essential to understanding and predicting the response of materials to various loading conditions and are fundamental to continuum mechanics. Constitutive conditions normally include a bunch of boundaries that portray the material properties, like versatility, pliancy, consistency, and warm extension. How the material deforms or responds to applied forces or temperature changes is determined by these parameters.

In the case of linear elasticity, for instance, the constitutive equation connects a material's stress, or internal force per unit area, to its strain, or deformation per unit length. Hooke's law, which states that the stress is proportional to the strain and that the material's Young's modulus serves as the proportionality constant, is frequently used to describe this relationship. Complicated constitutive equations are used for non-linear materials like plastic or viscoelastic materials. These conditions might consolidate extra material boundaries, for example, yield strength, strain solidifying, or a time-subordinate way of behaving.

Depending on the material and assumptions made, constitutive equations can take various forms. They can be represented by straightforward mathematical expressions like linear or nonlinear equations, or they can be more complex models like differential equations or integral formulations. In the end, constitutive equations provide a mathematical framework for describing the mechanical response of materials. This makes it possible for scientists and engineers to simulate and examine the behavior of structures and components under a variety of loadings and environmental conditions. By understanding the constitutive conditions overseeing a material, we can pursue informed choices in different fields, including mechanical design, materials science, structural design, and numerous others.

Yield Criteria

Yield standards, otherwise called yield surfaces or yield conditions, are numerical articulations or conditions that characterize when a material goes through plastic distortion under outside stacking. They can be used as a guide to figure out when a material goes from being elastic to being plastic. Yield rules are fundamental in designing and materials science as they give a system for foreseeing and dissecting the beginning of yielding in different materials and designs. To accommodate various material behaviors and loading conditions, there are various yield criteria. One broadly utilized measure is the Von Mises rule, which applies to pliable materials encountering isotropic plastic disfigurement. It expresses that yielding happens when the von Mises-comparable pressure surpasses a basic worth. The Tressa criterion, which focuses on significant shear stress and dictates yielding when the maximum shear stress exceeds a critical value, is another frequently used criterion.

The Mohr-Coulomb criterion is frequently applied to rocks and soils, which have cohesion and frictional properties. To determine yielding, it combines a linear tensile strength (cohesion) and a shear strength (friction) component. By taking into account the influence of hydrostatic pressure, the Drucker-Prager criterion enhances the Mohr-Coulomb criterion and is suitable for materials like concrete and granular soils. Hill's criterion is used when materials exhibit anisotropic behavior. The ability to predict yielding in anisotropic materials is made possible by Hill's criterion, which takes into account the strength in various orientations or directions. Structure and component design, analysis, and optimization all depend on yield criteria. Engineers can assess the structural integrity, anticipate plastic deformation, identify failure modes, and guarantee the safety and dependability of engineering systems by employing the appropriate yield criterion. The characteristics of the material, the loading conditions, and the level of accuracy required for a given application all influence the selection of a particular yield criterion.

TEMPERATURE DEPENDENCE OF FLOW STRESS

In the field of materials science and engineering, a significant phenomenon is the temperature dependence of flow stress. It refers to the temperature-dependent variation in flow stress or resistance to plastic deformation. For predicting the mechanical behavior of materials under various thermal conditions and optimizing their performance in various applications, it is essential to comprehend the temperature dependence of flow stress. In this writing audit, we will investigate key examinations and advancements connected with the temperature dependence of stream pressure, including the fundamental systems, trial procedures, numerical displaying approaches, and functional ramifications. Lattice vibration, dislocation motion, and thermally activated processes all have an impact on the temperature dependence of flow stress. At low temperatures, the stream pressure normally increases with diminishing temperature because of decreased cross-sectional vibrations and expanded protection from disengagement movement. This conduct is depicted by the warm actuation hypothesis, where disengagement movements require adequate nuclear power to conquer energy boundaries. Lattice vibrations become more pronounced as temperatures rise, facilitating dislocation motion and lowering flow stress. However, a phenomenon known as dynamic recovery or dynamic recrystallization can take place at high temperatures, resulting in a rapid softening of the material.

Trial strategies are utilized to quantify and portray the temperature dependence of stream pressure. Normal methods incorporate ductile testing, pressure testing, and hot distortion testing. To determine the temperature-dependent flow stress, tensile and compression tests are carried out at various temperatures. The material is subjected to deformation in hot deformation tests at temperatures that are typically higher than its recrystallization temperature. These tests consider the examination of the powerful mellowing conduct of materials at high temperatures. High-level methods, like high-temperature microscopy and in-situ mechanical testing, give bits of knowledge into the microstructural changes and deformity systems happening during temperature-subordinate distortion. The temperature dependence of flow stress is described and predicted using mathematical models. Experimental connections, like the Arrhenius condition and the power regulation condition, are regularly used to correlate stream pressure with temperature and other material boundaries. An exponential relationship is used in the Arrhenius equation to connect the activation energy, temperature, and flow stress. The power regulation condition, otherwise called the Holloman condition, addresses the stream pressure as a power-regulation capability of strain rate and temperature.

These models aid in predicting flow stress under various thermal conditions and provide a quantitative understanding of the temperature-dependent behavior of materials.

In a variety of fields, the temperature dependence of flow stress has significant practical implications. Understanding the temperature-dependent behavior of materials is essential for optimizing processing parameters like temperature and strain rate in materials processing and manufacturing to achieve desired mechanical properties and product characteristics. The design of structures for high-temperature applications, such as nuclear power plants or aerospace, is influenced by the temperature dependence of flow stress in structural engineering. It is essential for determining safe operating limits under various thermal conditions and assessing structural integrity. In addition, it is essential to know the temperature dependence of flow stress when choosing materials and designing them so that they can withstand the expected temperature changes and environmental conditions. A lot of research has been done on the underlying mechanisms, experimental methods, mathematical modeling approaches, and practical implications of the temperature dependence of flow stress. Understanding this peculiarity has empowered specialists and architects to streamline material handling, plan strong designs, and guarantee the solid presentation of materials and parts under various temperature conditions. Research in this field will additionally improve how we might interpret the temperature-subordinate way of behaving materials and work with the advancement of cutting-edge materials for a large number of utilizations.

Thermally Activated Mechanisms

The phenomenon in which the temperature influences the movement or transformation of atoms or molecules in a material is referred to as thermally activated mechanisms, thermally activated processes, or thermal activation. These systems assume a huge role in different physical and chemical processes, especially in strong-state materials. At the nuclear or sub-atomic level, warm actuation is driven by the energy given by warm vacillations. Thermal energy becomes sufficient to overcome energy barriers and facilitate atomic or molecular rearrangements at higher temperatures, resulting in changes in the behavior and properties of the material.

Diffusion, in which atoms or molecules move through a material, is one of the primary thermally activated mechanisms. Dissemination can happen through different components, like opening dispersion, interstitial dissemination, or grain-limit dispersion. As the temperature climbs, the energy accessible to iotas increments, advancing their dissemination and influencing the general material properties of vehicles. Solid-liquid and solid-solid phase transitions are two examples of phase transformations that are influenced by thermal activation. Thermal energy can cause atoms or molecules to rearrange themselves as temperature changes, resulting in structural changes and the formation or dissolution of phases.

Additionally, a variety of mechanical processes can be affected by mechanisms that are activated by heat. Thermally activated mechanisms, for instance, play a crucial role in materials that are subjected to creep, which is a time-dependent deformation that occurs under constant load or stress. The high temperature takes into account the development of disengagements inside the material, prompting plastic distortion over the long run. In the field of materials science and design, understanding and controlling thermally actuated components is imperative for planning materials with explicit properties and streamlining fabricating processes. By concentrating on the temperature dependence of these components, researchers and designers can anticipate material conduct under various warm circumstances and design materials for wanted applications. Furthermore, in fields like solid-state physics, catalysis, and chemical reactions, where the desired transformations are driven by temperature-induced changes, thermally activated processes are essential.

Effect of Temperature on Dislocation Behavior

Temperature significantly affects the behavior of disengagements, which are line surrenders in the precious stone cross-section of material. Dislocations play a crucial role in determining a material's strength, ductility, and deformation behavior in terms of its mechanical properties. The following is a summary of how temperature affects dislocation behavior:

- **Displacement Mobility:** Temperature influences the versatility of separations, which determines their capacity to travel through the gem cross-section. At higher temperatures, nuclear power increases, giving extra energy to disengagements and advancing their movement. Dislocations can move more easily as a result, increasing material ductility and plastic deformation.
- **Generation of Dislocation and Destruction:** The formation and destruction of dislocations within a material are influenced by temperature. Through mechanisms like thermal activation and climbing, thermal energy can encourage the formation of dislocations at high temperatures. Alternately, at lower temperatures, the nuclear power is diminished, making it more likely for disengagements to be produced or obliterated. This can influence the general thickness and appropriation of disengagements in the material.
- **Separation, Collaboration, and Sticking:** Temperature influences the connection among disengagements and different cross-

sectional deformities or hindrances present in the material. At higher temperatures, nuclear power can debilitate the communication among separations and obstructions, like pollution, hastens, or grain limits, permitting disengagements to effectively sidestep or defeat them more. On the other hand, at lower temperatures, nuclear power diminishes, reinforcing communication and making it more provoking for disengagements to move past hindrances. This can bring about expanded material strength and decreased pliability.

- **Cross-Slip Activation by Heat:** Cross-slip alludes to the development of separations starting with one slip plane and then onto the next inside the gem grid. At raised temperatures, nuclear power works with the initiation of cross-slip, permitting disengagements to change their coast planes all the more promptly. This can influence the general disengagement movement and the twisting way the material behaves.
- **Mechanisms for Deformation Activated by Heat:** Temperature can likewise impact the prevailing disfigurement systems in materials. For example, at high temperatures, disengagement float and climb components might rule, prompting plastic misshapen Ing and streaming. Notwithstanding, at lower temperatures, different components, for example, twinning or martensitic changes, may turn out to be more noticeable as temperature-subordinate stage advances happen.

Predicting and controlling the mechanical response of materials across various temperature regimes necessitates a thorough comprehension of the dislocations' temperature-dependent behavior. It aids in the design of materials for specific applications, the improvement of manufacturing procedures, and the monitoring of components' performance and structural integrity in response to a variety of temperature conditions.

Role of Grain Boundaries

The behavior and properties of polycrystalline materials are heavily influenced by grain boundaries. They are the points of interaction or limits between individual translucent locales, known as grains, inside a material. The overall mechanical, electrical, thermal, and chemical properties of a material can be significantly affected by the properties and behavior of grain boundaries. Here are a few critical jobs for grain limits: Dislocation motion can be impeded by grain boundaries, preventing dislocation propagation and plastic deformation. Material hardness and strength may rise as a result. Furthermore, grain limits can act as sources and sinks for disengagements, impacting their dissemination and influencing the general pliancy and flexibility of the material.

Under specific circumstances, grain limits can work with relative sliding or the development of nearby grains, known as grain limit sliding. Material ductility, particularly at high temperatures, may benefit from this mechanism's capacity to accommodate deformation. Grain limits can act as particular ways for the dissemination of iotas or particles inside a material. Grain growth, solid-state reactions, and diffusion-controlled phenomena can all benefit from the acceleration of mass transport provided by the presence of grain boundaries. In the processes of grain growth and recrystallization, grain boundaries play a crucial role. Changes in grain size and microstructure are brought about by the formation and migration of grain boundaries during these processes. Grain development can happen under warm or mechanical impacts, influencing material properties like strength, surface, and grain limit character.

Due to their distinct atomic arrangements from those of the grains, grain boundaries have the potential to impede the flow of thermal energy and electrical current. Grain limits might cause the dissipation of charge transporters or phonons, lessening the general electrical and thermal conductivity of the material. Materials' corrosion resistance and diffusion behavior can be affected by grain boundaries. Grain boundaries can sometimes serve as diffusion barriers or preferential locations for the onset of corrosion, preventing the passage of corrosive or diffusing species across the boundary. In materials science and engineering, controlling and understanding grain boundaries is very important. By altering the characteristics of the grain boundary, such as grain size, orientation, and boundary structure, scientists and engineers can modify the microstructure and properties of materials. Methods, for example, grain limit design and grain limit alteration, are utilized to upgrade material execution, work on mechanical properties, and improve material reaction under different circumstances.

EXPERIMENTAL METHODOLOGY

Throughout history, experimental methodology has been an essential component of scientific research and discovery. To gain a better understanding of the natural world, early civilizations used observations and experiments to develop experimental methodology. For instance, ancient Egyptians experimented with methods of embalming, and ancient Greeks like Aristotle and Archimedes experimented with a variety of scientific phenomena. Nonetheless, it was during the logical transformation in the sixteenth and seventeenth hundreds of years that exploratory strategy started to create an orderly way to deal with logical requests. Crafted by trailblazers, for example, Galileo Galilei and Francis Bacon established the groundwork for the cutting-edge logical technique, underscoring the significance of precise perception, trial and error, and the definition of speculations.

With the development of fields like physics, chemistry, and biology in the 18th and 19th centuries, experimental methodology made significant progress. Researchers like Isaac Newton, Antoine Lavoisier, and Charles Darwin directed earth-shattering tests that molded comprehension we might interpret the normal world. Exploratory contraptions and instruments were created, permitting researchers to gauge and examine peculiarities with more prominent accuracy. In the 20th century, technological advancements and interdisciplinary collaborations fuelled further advancements in experimental methodology. New horizons for scientific inquiry were opened by the development of experimental tools and techniques like electron microscopes, spectrometers, and particle accelerators. The fields of material science, science, science, and numerous others saw astounding improvement through exploratory examinations.

The experimental methodology has grown to include engineering, the social sciences, and even computer science in the last few decades. The trial configuration has become more modern, with scientists utilizing factual examination, control gatherings, and randomization to guarantee the dependability and legitimacy of their discoveries. Also, headways in information assortment and examination strategies, like the utilization of sensors, imaging advancements, and computational displaying, have additionally improved trial abilities. The component of innovation and scientific research today. Scientists can use it to test hypotheses, look into causal relationships, and produce empirical evidence to back up or refute theories. The behavior of natural and engineered systems can be better understood through experimental studies, which aid in the development of new technologies, the resolution of challenging issues, and a deeper comprehension of the world around us.

It has had a significant impact on a wide range of scientific discoveries and advancements. Researchers have been able to gain a deeper comprehension of the natural world and contribute to the advancement of human knowledge through the continuous improvement and application of experimental methods. The experimental methodology will undoubtedly continue to be an essential tool for scientific inquiry and exploration even as technology continues to advance.

Material Selection

Material determination is a pivotal part of designing and configuration processes, including the deliberate assessment and choice of materials in light of their properties, execution, and reasonableness for explicit applications. A product's or structure's functionality, dependability, and durability are directly influenced by the materials used. It involves taking into account a variety of aspects, including cost, availability, environmental considerations, electrical conductivity, corrosion resistance, and thermal properties. In this context, a methodical approach to material selection includes defining requirements, evaluating available materials, and making educated choices based on technical and financial considerations. Understanding the application's or project's requirements and constraints is the first step in the material selection process. This entails specifying the functional requirements, load conditions, operating environment, anticipated lifespan, and any applicable regulatory or industry standards. Engineers can establish a baseline for selecting materials that will perform optimally under the conditions by clearly defining these parameters.

The next step is to identify and evaluate potential material options after the requirements have been defined. This entails conducting research and compiling data on various materials with the desired properties and performance characteristics. When looking for suitable materials, material databases, technical literature, and expert knowledge are all useful sources of information. Materials are typically evaluated by comparing their properties and performance to predetermined requirements. Material testing, data analysis, and simulations are all ways to accomplish this. For instance, information regarding tensile strength, hardness, impact resistance, and other mechanical properties can be obtained through mechanical testing. Thermal analysis methods can also evaluate materials' heat resistance, expansion coefficient, and thermal conductivity. Consumption tests, electrical conductivity tests, and other particular tests can likewise be used to assess explicit properties.

During the selection process, in addition to material properties, costs, availability, and processability must also be taken into account. The material's cost per unit and its overall impact on the project budget are two areas of cost consideration. The material's accessibility and availability in the required quantities and timeframes are referred to as its availability. The material's machinability, weldability, and compatibility with production methods are all taken into consideration when determining its processability—the ease with which it can be manufactured, shaped, and joined. Natural contemplations are progressively significant in material choice cycles. To minimize the product's or structure's impact on the environment throughout its life cycle, environmentally friendly and sustainable materials are given preference. Factors like recyclability, energy utilization during assembly, and the potential for reusing or discarding the material after use are considered.

The best match between the material properties, performance requirements, cost, availability, and environmental considerations is the basis for the final decision after the evaluation is completed. Because materials may be superior in some properties but inferior in others, this choice frequently involves making trade-offs. The picked material is then incorporated into the plan and design interaction, taking into account factors like material similarity with different parts, get-together prerequisites, and long-haul upkeep contemplations. Material choice is a basic cycle in designing and configuration, impacting the exhibition, usefulness, and strength of an item or construction. It involves a methodical approach to defining requirements, evaluating available materials, and making educated choices based on technical, financial, and environmental factors. Engineers can ensure that their designs achieve the desired performance goals and produce the best quality, dependability, and sustainability outcomes by carefully selecting the appropriate materials.

Specimen Preparation

Example readiness is a fundamental stage in directing trials and tests on materials. It involves making samples or specimens from the chosen material with care and precision. The specific requirements of the test, such as tensile, compression, or fatigue testing, determine the preparation process. It might include cutting, machining, crushing, or projecting the material to acquire the ideal shape and aspects. As part of specimen preparation, the samples' surface quality and cleanliness are checked, and any necessary surface treatments or coatings are applied. Appropriate example planning is essential to getting exact and dependable test results, as the quality and consistency of the examples directly influence the result of material portrayal and execution assessment.

Mechanical Testing

The evaluation of materials' mechanical properties and behavior relies heavily on mechanical testing. Mechanical testing provides useful insights into a material's strength, durability, elasticity, and other significant characteristics by subjecting it to controlled loads and deformations. The purpose of this paragraph is to examine the significance, types, and parameters of mechanical testing, among other aspects. The basic role of mechanical testing is to determine the way that materials will perform under various stacking conditions, assisting architects and researchers with pursuing informed choices concerning material determination and planning. A variety of tests are used to evaluate mechanical properties like tensile strength, yield strength, ductility, hardness, and fatigue resistance.

One of the most common mechanical methods for determining a material's response to tension is tensile testing. An example is a hub that is exposed to a rising hub load until it breaks. The subsequent pressure strain bend gives significant data about the material's versatile and plastic way of behaving, extreme strength and stretching before disappointment. Pressure testing, then again, includes exposing a material to compressive force until it misshapes or breaks. Because it helps determine a material's capacity to withstand compressive loads, this test is especially crucial for structural materials.

The flexural strength and stiffness of the material are evaluated through bending tests. In these tests, an example is exposed to a three-point or four-point bowing setup, re-enacting certifiable situations where materials experience twisting burdens. The experimental outcomes help in the planning of primary parts, like bars or segments. A material's capacity to absorb energy under sudden, dynamic loading conditions is evaluated through impact testing. It involves striking a sample with a pendulum or a falling weight to measure the amount of energy absorbed and the behavior of the fracture that results. The sample can be notch-less or unbottled. Influence testing is vital for materials utilized in applications where abrupt loads or effects are normal, like car parts or defensive gear.

The material's resistance to localized plastic deformation or indentation is determined through hardness testing. Based on the properties of the material and the accuracy that is desired, a variety of hardness tests, such as the Rockwell, Vickers, or Brinell tests, are used. The wear resistance, strength, and suitability of a material for particular applications are frequently assessed using hardness values. Weariness testing inspects a material's capacity to endure cyclic stacking and repeated pressure cycles. In structures that are subjected to varying loads, the material breakdown is frequently brought on by fatigue failure. Engineers can determine a material's fatigue life and safe operating limits by repeatedly loading a sample and monitoring its response.

In addition to these tests, numerous specialized mechanical tests are tailored to specific materials and applications. For example, creep testing looks at how a material deforms over a long period under constant load to determine its long-term stability. Similarly, for components that are subjected to torque, such as springs or shafts, torsion testing is essential because it measures a material's resistance to twisting forces. To guarantee results that are both accurate and repeatable, mechanical testing requires the use of precise instruments and adherence to established protocols. Test examples are painstakingly prepared as per the laid-out rules, and test boundaries, for example, load, deformity rate, and temperature, are controlled and checked all through the tests. Digital imaging, strain gauges, and extensometers are among the cutting-edge tools used to record precise measurements and track the material's behavior throughout the testing process. Engineers can evaluate parameters like strength, elasticity, durability, and fatigue resistance through a variety of tests that place materials under controlled loads and deformations. Material selection, design, and ensuring the safety and dependability of structures and components in a variety of industries are all aided by the findings of mechanical testing.

Temperature Control

It plays a crucial role in a wide range of applications and industries, including food processing, environmental control systems, and scientific research. Maintaining a comfortable working environment, preserving materials, ensuring safety, and achieving optimal product quality all require efficient temperature control. This section will investigate the significance of temperature control, various strategies for temperature control, and its applications in different fields. In industries where precise and stable temperatures are required to achieve desired outcomes, temperature control is essential. In assembling, temperature control is important to guarantee predictable item quality and to work with proficient creation processes. For instance, for microchips to attain the desired properties and performance, the semiconductor industry must ensure that precise temperatures are maintained during the fabrication process. Similarly, maintaining the stability and efficacy of vaccines and drugs in the pharmaceutical industry necessitates temperature control.

Temperature control is essential for ensuring food safety and extending the shelf life of perishable goods during food processing and storage. The growth of bacteria and other microorganisms that are capable of causing foodborne illness is inhibited by proper temperature control. Additionally, temperature control is essential for preserving food products' sensory qualities and nutritional value. Environmental control systems like heating, ventilation, and air conditioning (HVAC) also require temperature control. To provide occupants with a comfortable and productive environment, these systems control the temperature of indoor spaces. Successful temperature control upgrades solace as well as energy proficiency by enhancing warming and cooling processes in light of inhabitancy and outer circumstances.

There are a few techniques and innovations utilized for temperature control. Utilizing thermostats and temperature sensors, which monitor the surrounding temperature and activate heating or cooling systems as necessary to maintain the desired set point, is one common strategy. HVAC systems, both residential and commercial, frequently employ this approach. In modern cycles, further developed temperature control techniques are utilized. Temperature control, for instance, in refrigeration systems is accomplished by precisely controlling the flow of refrigerant and the operation of the compressor. The control of fluid flow rates and heat transfer mechanisms is how temperature control is achieved in thermal management systems like cooling towers or heat exchangers.

Temperature control is frequently accomplished with specialized tools like incubators, ovens, or environmental chambers in research facilities and laboratories. These gadgets give exact temperature control inside a controlled climate, permitting scientists to direct investigations or store tests under unambiguous temperature conditions. In numerous applications and industries, temperature control is critical. In addition to creating a cozy indoor environment, it ensures product quality, safety, and efficiency in the manufacturing, food processing, and storage processes. Temperature control can be achieved using a variety of technologies and methods, including thermostats and sensors in HVAC systems, advanced industrial systems, and specialized laboratory equipment. No matter what the application, exact temperature control is fundamental for accomplishing desired results, keeping up with the honesty of materials, and guaranteeing ideal execution and security.

RESULTS AND DISCUSSION

Experimental Data Analysis

In scientific and engineering research, experimental data analysis is a crucial step in which collected data are examined, processed, and interpreted to find relevant information and draw conclusions. To discover relationships, trends, and patterns in the data, this procedure makes use of a variety of statistical and analytical methods. In this passage, we will investigate the significance of trial information examination, the normal methods utilized, and the meaning of precise information translation. There are numerous reasons why experimental data analysis is important. First, it lets researchers compare experimental results to predicted outcomes to validate hypotheses and test the accuracy of theoretical models. Researchers can gain insight into the validity and applicability of their theories by determining the extent to which their hypotheses align with the observed data through the process of data analysis.

Additionally, experimental data analysis contributes to the identification and comprehension of the underlying factors and mechanisms that influence the observed phenomena. By looking at the connections between various factors, specialists can acquire a more profound comprehension of the perplexing communications and conditions inside the framework being scrutinized. Improved models, enhanced designs, and more effective procedures may result from this knowledge. Depending on the nature of the data and the goals of the research, experimental data analysis uses a variety of methods. A summary of the data's variability and descriptive statistics, such as mean, standard deviation, and range, are provided. The central tendencies and distribution of the data points are first revealed by these statistics.

Based on a sample, inferential statistics are used to make inferences or draw conclusions about a population. Researchers can determine the statistical significance of their findings and the reliability of their conclusions with the assistance of tools like hypothesis testing and confidence intervals. Another common method is regression analysis, which looks at how variables are related to one another. This makes it possible to estimate parameters and find trends. Perception techniques, like diagrams, graphs, and plots, are integral assets for information examination as they give a visual portrayal of the information designs. Data relationships, distributions, and variations are frequently shown using scatter plots, line graphs, histograms, and box plots. Outliers, trends, and anomalies can be more easily identified using these visual representations, which makes it easier to interpret and communicate the results.

For valid conclusions to be drawn, experimental data must be accurately interpreted. The uncertainties and limitations of their experiments and data collection strategies must be taken into account by researchers. Factors like estimation blunders, inspection predisposition, and perplexing factors can affect the dependability and generalizability of the outcomes. To mitigate these issues and ensure the validity of the data-derived conclusions, proper statistical analysis and critical evaluation are required. All in all, trial information examination is a basic move toward logical exploration and design. It permits specialists to approve speculations, figure out complex connections, and reach solid inferences. Researchers can extract meaningful information from the data, identify patterns and trends, and make informed decisions by utilizing statistical and analytical methods. Considering the experimental process's limitations and uncertainties, accurate data interpretation is crucial. Generally speaking, exploratory information examination is a useful asset for propelling information and driving development in different fields.

Flow Stress-Temperature Relationships

Stream pressure temperature connections allude to the reliance of stream pressure, which is the protection from plastic distortion, on temperature. Predicting and analyzing the mechanical behavior of materials across a wide temperature range requires an understanding of the relationship between flow stress and temperature. The material's composition, microstructure, and deformation mechanisms all have an impact on the flow stress-temperature relationship's overall complex behavior. The temperature dependence of stream pressure can be portrayed by various areas or systems, each represented by unambiguous components. The flow stress typically rises with the temperature at low temperatures. This behavior is attributed to diminished warm actuation of disengagement movement and expanded protection from plastic misshapen Ing. At low temperatures, the portability of separations is restricted, and the material turns out to be more inclined to disengage heap-ups and solidify systems. As the temperature drops, this causes flow stress to rise.

The flow stress tends to decrease as the temperature rises. Increased thermal activation of dislocation motion is primarily to blame for this. At higher temperatures, the nuclear power accessible to the material works with disengagement float and diminishes the protection from plastic distortion. As the temperature rises, the material shows more ductility and less flow stress. Be that as it may, at raised temperatures, dynamic recuperation and dynamic recrystallization can happen, prompting a quick relaxation of the material. Dynamic recuperation includes the rebuilding of separation structures, decreasing their thickness, and expanding the material's malleability. Dynamic recrystallization, then again, brings about the arrangement of new, sans-strain grains inside the material. These peculiarities can essentially impact the stream pressure-temperature relationship and result in a sharp reduction in stream pressure at higher temperatures.

The stream pressure-temperature relationship is often depicted using exact models or numerical conditions. The Arrhenius condition is a commonly utilized model that relates the stream pressure to the initiation energy and temperature. It explains the exponential relationship between the reciprocal of the absolute temperature and the flow stress. The rate of temperature-dependent variation in flow stress is determined by the activation energy, which represents the energy barrier necessary for dislocation motion. The relationship between flow stress and temperature in particular materials has been proposed to be explained by physical models and theories based on the underlying deformation mechanisms, in addition to empirical models. Thermal activation theory and dislocation dynamics models, for instance, shed light on the temperature-dependent behavior of flow stress in metals. The relationship between flow stress and temperature is taken into account by these models, as are the effects of dislocation interactions, thermal energy, and dislocation density.

Numerous engineering applications are significantly impacted by the flow stress-temperature relationship. Because materials need to exhibit the desired mechanical properties under operating conditions, this influences the selection of materials for particular temperature ranges. Predicting material behavior during manufacturing processes like hot forming and heat treatment is also made easier by having an understanding of the relationship between flow stress and temperature. In addition, it is essential in the design of structures and parts that are subject to temperature changes, ensuring their dependability and preventing failure. The composition, microstructure, and deformation mechanisms of the material all play a role in the intricate flow stress-temperature relationship. It exhibits distinct regimes, with flow stress decreasing at higher temperatures and increasing at lower temperatures. At high temperatures, the occurrence of dynamic recovery and dynamic recrystallization can result in rapid softening. The flow stress-temperature relationship is described by empirical models like the Arrhenius equation and physical models based on deformation mechanisms. Material selection, process optimization, and the design of structures that can withstand temperature variations all depend on this relationship.

Comparison with Theoretical Models

Comparing experimental data to theoretical models of flow stress-temperature relationships is common when analyzing experimental data. To predict and interpret experimental data, theoretical models provide a framework for comprehending the underlying mechanisms and behaviors of materials. For various materials, flow stress-temperature relationships are described by several theoretical models. Dislocation theory, crystal plasticity, or deformation mechanisms specific to particular materials are common foundations for these models in material science.

Researchers can assess the validity of the models and gain insight into the predominant mechanisms governing the behavior of the material by comparing experimental data with theoretical models. Assuming the exploratory information adjusts well to the expectations of a specific model, it is recommended that the model precisely captures the basic physical science and can be utilized for future expectations or designing applications. On the other hand, differences between experimental data and theoretical models could be a sign of the model's limitations or simplifications, or they could be a sign that additional factors exist that the model does not take into account. The theoretical framework's accuracy and applicability may be enhanced by further investigation and refinement as a result of these discrepancies.

Experimental data that significantly deviates from theoretical models may indicate the need for new models or modifications to existing models in some instances. To develop more comprehensive models that better convey the observed behavior, researchers may investigate alternative approaches, take into account various deformation mechanisms, or incorporate additional variables. It is essential to keep in mind that theoretical models are merely simplified versions of intricate reality and frequently assume idealized conditions. As a result, it is not always expected or required that experimental data and theoretical predictions coincide completely. The search for models with reasonable approximations and insights into the flow stress-temperature relationship is the objective, as these models will serve as a guide for material design, process optimization, and structural analysis.

A useful practice in the analysis of flow stress-temperature relationships is comparing experimental data with theoretical models. It makes it possible for researchers to validate and improve existing models, discover their limitations, and motivate additional research. To interpret and predict the outcomes of experiments, theoretical models provide a framework for comprehending the fundamental mechanisms and behaviors of materials. Researchers can improve their understanding of material behavior and material performance in a variety of applications by iteratively improving experimental data-based models.

MODELING APPROACHES

Phenomenological Models

Phenomenological models are mathematical descriptions that aim to capture and represent the observed behavior of a system or phenomenon. They can be empirical or semi-empirical. These models depend on trial information and can't be guaranteed to consider the basic actual components administering the peculiarity. Phenomenological models' characteristics and applications will be examined in this section. Phenomenological models are constructed without explicitly incorporating the fundamental physical principles or mechanisms that underlie the observed behavior. Instead, they are based on the analysis and interpretation of experimental data. Without necessarily providing insights into the processes that lie beneath, these models concentrate on describing the connection that exists between the output response and the input variables. Phenomenological models frequently include fitting numerical conditions or working with trial information utilizing relapse investigation or enhancement methods. The data's observed trends and patterns typically serve as a guide for the model's mathematical form selection. The next step is to minimize the differences between the predictions of the model and the experimental data to determine the parameters of the model.

Phenomenological models are frequently used in material science and engineering to describe material behavior like stress-strain relationships, creep deformation, fatigue life, and thermal expansion. These models furnish engineers with a numerical system to foresee and dissect the material's reaction under various stacking and natural circumstances, empowering the planning and improvement of designs and parts. While phenomenological models enjoy their benefits as far as effortlessness and reasonableness go, they also have constraints. Since they depend exclusively on observational information, they may not have experience with the basic instruments or actual cycles. Outside of the range of the experimental data used in the development of the model, these models also have limited extrapolation capabilities. Therefore, phenomenological models should be used with caution when applied to unexplored areas or conditions.

Mechanistic Models

In contrast to phenomenological models, mechanistic models aim to explicitly consider the fundamental physical mechanisms and principles governing behavior to describe a system or phenomenon. These models frequently involve intricate mathematical formulations or simulations and are founded on a thorough comprehension of the underlying physics and mathematics. In this section, we will investigate the qualities and utilization of robotic models. The fundamental rules and regulations that govern the system under study serve as the foundation for mechanistic models. Conservation laws, constitutive relationships, and governing equations are just a few of the physical processes that need to be thoroughly comprehended to perform them. By explicitly taking into account the system's interactions, forces, and energy transfers, these models aim to accurately depict the system's behavior.

Mathematical equations or computational algorithms based on fundamental physics and mathematics are frequently used in mechanistic models. The physical mechanisms involved are taken into account when these equations describe the relationships between the various variables and parameters. Material properties, experimental measurements, or calibration processes typically provide the model parameters. Numerous areas of engineering and science make use of mechanistic models. Fluid dynamics, electromagnetic fields, and quantum mechanics are all examples of complex physical systems that are described by mechanistic models in physics. New insights and discoveries are made possible by these models, which enable researchers to comprehend and anticipate the system's response under various conditions.

Molecular Dynamics Simulations

Sub-atomic element recreations are computational techniques used to concentrate on the powerful way iotas and particles behave over the long run. The motion and interactions between individual atoms are modeled and simulated in these simulations using either classical or quantum mechanical principles. In the following paragraph, we will investigate the fundamentals of molecular dynamics simulations, their applications, and their significance in a variety of scientific fields. The idea of force fields, which describe the interatomic forces and potential energy surfaces, is at the center of molecular dynamics simulations. In classical MD simulations, the interactions between atoms are represented by simplified mathematical models, such as the Lennard-Jones potential or the Coulombic potential. Quantum MD recreations, then again, utilize quantum mechanical standards, for example, thickness practical hypotheses or wavefunction-based strategies, to display the electronic design and connections.

Numerical integration methods like the Varlet algorithm and the leapfrog method are used to iteratively calculate the positions, velocities, and forces acting on each atom in MD simulations. By coordinating Newton's laws of movement, the direction of every molecule still up in the air is preserved, permitting the reproduction of the framework's conduct over the long run. Sub-atomic element recreations have a great many applications in different logical disciplines. They are utilized in the study of chemical reactions, the properties of materials, and molecular dynamics and kinetics in chemistry. Understanding the behavior of proteins, nucleic acids, and other biomolecules, as well as drug discovery, protein folding studies, and molecular biology, has been made possible by MD simulations.

The study of the mechanical properties of materials, such as strength, elasticity, and fracture behavior, is carried out in materials science using molecular dynamics simulations. The design and optimization of new materials with desired properties are made possible by these simulations, which shed light on the atomic-scale processes that control material behavior. In nanoscience and nanotechnology, MD simulations are vital for understanding and controlling nanoscale frameworks. Insights into the thermal, mechanical, and electrical properties of nanoparticles, nanotubes, and nanocomposites can be gained from their ability to simulate their behavior. The creation of customized nanoscale devices, sensors, and materials is aided by these simulations.

Atomic element recreations likewise track down applications in physical science, topography, and natural science. They are used to investigate phenomena like heat transfer, phase transitions, and the behavior of liquids, gases, and plasmas. MD simulations are utilized in the research of protein-protein interactions, protein-ligand interactions, and the dynamics of biomolecular systems in computational biology. The ability of molecular dynamics simulations to provide atomic-level details and insights that are frequently difficult or impossible to obtain experimentally is one of their main benefits. They permit researchers to notice and break down the elements of individual iotas and particles, giving them a more profound comprehension of the basic cycles. Before undertaking costly and time-consuming experimental studies, MD simulations also provide a cost-effective and time-efficient method for investigating and predicting the behavior of complex systems.

APPLICATION IN ENGINEERING

Engineering uses flow stress-temperature relationships more and more as a result of the need to comprehend and anticipate the mechanical behavior of materials under various thermal conditions. The early developments in materials science and the industrial revolution can be traced back to the application of these relationships in engineering. During the modern unrest in the eighteenth and nineteenth centuries, there was a developing interest in further and more solid materials to help the progressions in hardware, foundation, and transportation. Scientists and engineers began to understand how temperature affects materials' mechanical properties and suitability for various applications.

One prominent authentic model is the advancement of steam motors. The significance of temperature and pressure in the operation of steam engines was recognized by engineers like James Watt and George Stephenson. They had to choose materials that wouldn't deform or break too much under high temperatures and pressures. They were able to optimize the design and material selection for steam engine components by comprehending the flow stress-temperature relationships of materials. This resulted in improved performance and efficiency. In the mid-twentieth century, progressions in metallurgy and the advancement of new combinations further highlighted the significance of stream pressure and temperature connections. Engineers started to concentrate on the way metals behave at raised temperatures, especially in applications such as airplane motors and power plants. The selection of materials for high-temperature environments required an understanding of flow stress-temperature relationships, which ensured the safety and dependability of these systems.

In the second half of the 20th century, computational modeling and simulation methods made it easier for engineers to accurately predict how materials would behave at different temperatures. Virtual testing and design optimization were made possible by incorporating flow stress-temperature relationships into FEA and other simulation tools. This capacity upset the designing field, empowering specialists to assess the exhibition of parts and designs practically and effectively. The range of applications for flow stress-temperature relationships has grown significantly over the past few decades as a result of advancements in materials science and engineering. High-temperature alloys, ceramics, and composites, among other advanced materials, have extended temperature capabilities in a variety of sectors. By understanding the stream pressure and temperature connections of these materials, architects can plan parts that can endure outrageous temperature conditions while keeping up with their underlying trustworthiness and execution.

The engineering fields of aerospace, automotive, energy, manufacturing, and structural engineering all benefit from the application of flow stress-temperature relationships. When designing turbine blades, engine components, and spacecraft structures that operate in high-temperature environments, for instance, in the aerospace industry, the flow stress-temperature relationships of materials are essential. In the field of automotive engineering, engine components, exhaust systems, and brake systems must take into account how materials behave at high temperatures. The evaluation of the behavior of materials in buildings, bridges, and other structures subjected to varying temperature conditions is carried out in structural engineering using flow stress-temperature relationships. The comprehension of these connections assists engineers with deciding safe working cut-off points, foreseeing underlying misshapen Ings, and guaranteeing the drawn-out respectability of the designs.

Engineering has used flow stress-temperature relationships for several centuries. This practise is very old. Engineers have relied on these relationships to select materials, improve designs, and guarantee the safety and reliability of various engineering systems since the industrial revolution to the present day, thanks to advancements in computational modeling and materials science. The development of materials and structures that are both more effective and more durable is made possible by the ever-increasing knowledge of flow stress-temperature relationships, which continues to drive engineering innovation and advancement.

Material Selection and Design

In a variety of industries and engineering fields, material selection and design are crucial. The selection of materials influences the presentation, dependability, and viability of an item or framework. In this section, we will investigate the significance of material choice and configuration, factors impacting the determination cycle, and procedures for streamlining material properties. Choosing the best material for a given application based on its properties, characteristics, and performance requirements is the process of material selection. It necessitates taking into account a variety of aspects, including cost, environmental impact, durability, manufacturability, thermal properties, chemical resistance, and mechanical properties. Material selection is incorporated into the design process by taking into account the desired functionality and performance of the product or system concerning the properties and capabilities of various materials. It includes coordinating the picked material into the general plan and considering variables like burden-bearing limit, underlying trustworthiness, layered dependability, and similarity with different parts or materials.

The material should meet the particular practical requirements of the application, like strength, solidity, conductivity, or erosion resistance. When choosing a material that can serve the intended purpose, it is essential to comprehend the performance requirements and constraints. Material selection is influenced by the operating environment, which includes temperature, humidity, chemical exposure, and mechanical stress. Choose materials that can withstand these conditions without significantly deteriorating or losing performance. The available manufacturing techniques and their compatibility with the chosen material influence the selection of materials. Processing methods like casting, forging, machining, or additive manufacturing may be required for various materials.

When choosing a material, an important consideration is the cost of the materials, including the costs of raw materials and processing. Offsetting execution necessities with cost imperatives is fundamental for a practical plan. Engineers can understand and evaluate the performance of various materials under specific conditions by using experimental testing and characterization techniques, which can provide valuable data on material properties. Material behavior and performance can be predicted and improved through the use of numerical models and simulations, such as finite element analysis and molecular dynamics simulations. These apparatuses help in the assessment of various material choices and the planning of creative arrangements. Material properties can frequently be improved through alteration methods, for example, alloying, heat treatment, surface medicines, or coatings. These cycles can work on mechanical strength, hardness, erosion resistance, or other desired attributes.

Progressions in materials science and design continually present new materials or variations with further developed properties. Investigating these choices and taking into account material replacements can prompt inventive plans and further development of execution. Product design and material selection are crucial aspects of engineering. Picking suitable materials because of useful prerequisites, natural circumstances, fabrication cycles, and cost considerations is fundamental for accomplishing desired execution and dependability. Testing, modeling, and material modification are all examples of optimization strategies that can support novel design solutions and further enhance material properties. By cautiously considering material determination and configuration, designers can foster items and frameworks that meet exhibition objectives, improve maintainability, and drive mechanical progress.

Process Optimization

A methodical approach to increasing manufacturing or operational processes' efficiency, productivity, quality, and cost-effectiveness is known as process optimization. It includes dissecting and calibrating different interaction boundaries and factors to accomplish ideal execution and desired results. Organizations can reduce waste, increase competitiveness, and drive overall business success by optimizing processes. We will discuss the significance of process optimization, common optimization methods, and their applications across a variety of industries in the following paragraph. Organizations that want to increase their operational efficacy and competitiveness need to implement process optimization. By smoothing out processes, associations can lessen costs, further develop efficiency, upgrade item quality, and increase consumer loyalty. Streamlining permits organizations to distinguish and wipe out bottlenecks, limit waste, and take full advantage of assets. It empowers associations to satisfy client needs more effectively, respond to market changes, and gain an upper hand in their particular ventures.

There are a few strategies and techniques usually utilized for process improvement. Structured approaches to systematic process analysis and improvement are provided by these methods. A portion of the normal enhancement methods include:

- **Six Senses:** The goal of the data-driven Six Sigma methodology is to eliminate defects and variations in processes. A methodical approach to defining, measuring, analyzing, enhancing, and controlling processes is required. To pinpoint and eliminate the underlying causes of process variations, Six Sigma employs statistical methods and tools, resulting in enhanced process quality and performance.
- **Lean Production:** Lean assembly centers around wiping out squander and non-esteem-added exercises from the creative interaction. By minimizing inventory, shortening wait times, reducing defects, and making the best use of resources, it aims to create a smooth and effective workflow. Lean devices, for example, esteem stream planning, 5S, Kaizen, and just-in-time (JIT) creation, are normally used to upgrade processes and accomplish functional greatness.

  SPC, or statistical process control, includes the utilization of factual procedures to screen and control process factors. It distinguishes process varieties, patterns, and irregularities, permitting associations to make convenient remedial moves. SPC empowers process soundness, diminishes changeability, and guarantees steady item quality.

- **Design of Examinations (DOE):** DOE is a planned method for systematically changing process parameters to see how they affect the output of the process. It assists businesses in determining the ideal process conditions for achieving their goals, comprehending the interactions between variables, and optimizing the settings of their processes. DOE permits associations to lead tests in a controlled way, break down the outcomes, and settle on information-driven choices for process improvement.
- **Engineering Value:** The goal of value engineering is to cut costs while still delivering the most value from a process or product. It entails examining a process's functions and requirements and locating opportunities for enhancement. Processes are optimized through the elimination of redundant steps, reduction in complexity, enhancement of efficiency, and identification of cost-effective alternatives through value engineering.

Structural Integrity Assessment

A crucial step in engineering and design is structural integrity assessment, which examines a structure's capacity to withstand loads and environmental conditions while maintaining its intended performance and safety. It includes a variety of approaches for evaluating the structural integrity of systems and components to make sure they meet design requirements and regulatory standards. In this paragraph, we will discuss the significance of structural integrity assessment, common assessment methods, and their applications in various industries. The evaluation of underlying trustworthiness is fundamental to guaranteeing the well-being and unwavering quality of designs across a great many enterprises, including aviation, structural design, oil and gas, auto, and assembly. Engineers can identify potential flaws, defects, or failures that could jeopardize the structure's performance or safety by carrying out structural integrity assessments. These evaluations assist in making well-informed decisions regarding strategies for replacement, repair, and maintenance to avoid catastrophic failures and guarantee the structure's longevity.

NDT methods make it possible to inspect and evaluate structures without compromising the system's or component's integrity. Ultrasonic testing, radiography, magnetic particle testing, and eddy current testing are all used to find defects, cracks, or material anomalies on the surface or subsurface that could compromise the structural integrity of the structure. The computational approach known as FEA is utilized to simulate and examine the behavior of structures under various loading conditions. Predicting stress distribution, deformation, and failure modes requires breaking up the structure into its parts using mathematical models. FEA assists in guiding design modifications or structural reinforcements by identifying areas of high stress or potential failure points.

In load testing, the structure is put through controlled and calibrated loads to see how it responds and behaves. In bridge testing, where the structure is loaded to determine its capacity, deflection, and structural response, this method is frequently employed. The results of load testing can be used to assess the structure's structural integrity and determine whether or not it satisfies safety and design standards. Using sensors and data acquisition systems, SHM is a continuous monitoring method that detects any changes or anomalies in real-time and monitors structural performance. SHM enables proactive maintenance and repair actions to prevent failures by enabling the early detection of structural damage, fatigue, or degradation. In the aeronautic trade, underlying honesty in evaluation is significant for guaranteeing the security and unwavering quality of airplane structures. Non-destructive testing of aircraft structures for cracks, corrosion, or material degradation is part of this process, as is evaluating critical components' fatigue life and inspecting composite materials for damage or delamination.

In civil engineering, structural integrity assessments are crucial for determining the safety and durability of structures like bridges, dams, and buildings. The structural stability, load-carrying capacity, and resistance to environmental factors like wind, corrosion, and earthquakes are all evaluated. The long-term performance and integrity of structures are also monitored with the help of structural health monitoring systems. Offshore platforms, pipelines, and pressure vessels require structural integrity assessments in the oil and gas industry. It involves assessing these assets' structural integrity in harsh environments like high-pressure, high-temperature environments, and corrosive substances. Evaluations are conducted to recognize imperfections, consumption, or weakness and guarantee consistency with security guidelines. Underlying trustworthiness appraisal is critical in auto and assembling enterprises to guarantee the security and dependability of vehicles, hardware, and modern designs.

LIMITATIONS AND FUTURE DIRECTIONS

The pliancy hypothesis and the temperature dependence of stream pressure in strong mechanics have enormously added to how we might interpret material ways of behaving and the plan of designing designs. However, there are still some restrictions and areas that require additional investigation. The complicated relationship between temperature and flow stress is one limitation. Although a lot of empirical models and mathematical equations, like the Arrhenius equation, have been used to describe this relationship, it's possible that they don't fully convey the intricate physical mechanisms at the heart of the phenomenon. Microstructural changes, dislocation dynamics, and thermally activated processes all have an impact on flow stress's temperature dependence. The development of more precise and comprehensive models that can take into account these intricate interactions and provide a deeper comprehension of the temperature dependence of flow stress ought to be the primary focus of any subsequent research.

The requirement for experimental data across a broad temperature and strain rate range to validate and improve existing models is another limitation. It can be hard to get accurate and reliable data, especially when working with new materials or in extreme temperatures. Therefore, the goal of any subsequent experimental studies ought to be to fill in these voids and supply a more extensive dataset for material characterization and model validation. Additionally, isotropic materials are the primary focus of the existing models and theories. In any case, many design applications include anisotropic materials, like composites or finished metals, where the stream pressure-temperature relationship can show more mind-boggling ways of behaving. The temperature dependence of flow stress in anisotropic materials and the development of models capable of capturing their distinct characteristics should be the focus of future research.

In addition, there is still a lack of knowledge regarding the temperature dependence of flow stress in complex material systems, such as nanomaterials or materials with microstructural heterogeneities. These materials frequently exhibit behavior that is size-dependent and interface-dominated, which can significantly alter the relationship between flow stress and temperature. Future investigations ought to dig into these intricacies and foster speculations and models customized to these specific material frameworks. The incorporation of methods for multiscale modeling is an additional direction for upcoming research. The way materials behave at the perceptible scale is affected by processes happening at the microstructural level. By capturing the interactions between dislocations, grain boundaries, and other microstructural features, integrating multiscale models can provide a complete comprehension of the temperature dependence of flow stress. Particularly for materials with intricate microstructures, this may improve the models' precision and ability to make predictions.

Lastly, numerical simulation methods for analyzing the temperature dependence of flow stress have the potential to be further refined and developed as a result of the rapid development of computational methods and high-performance computing. Crystal plasticity finite element models and molecular dynamics simulations are two examples of advanced modeling methods that can provide useful insights into the temperature-dependent behavior of materials at the atomic and mesoscales. The accuracy of flow stress-temperature predictions can be improved by combining these advanced modeling methods with experimental data. This can help us gain a deeper comprehension of the mechanisms that lie beneath the surface. While the pliancy hypothesis and the temperature reliance of stream pressure have essentially added to how we might interpret material ways of behaving, some impediments should be tended to. The development of more precise models, their validation with extensive experimental data, the investigation of the behavior of anisotropic and complex material systems, the incorporation of multiscale modeling strategies, and the utilization of cutting-edge computational techniques should all be the goals of subsequent research. By tending to these limits and propelling our insight here, we can additionally upgrade our capacity to foresee and enhance the mechanical way of behaving materials across a large number of temperatures and work on the plan and execution of designing designs.

Assumptions and Simplifications

In a variety of fields, including the sciences, engineering, economics, and social sciences, assumptions and simplifications are essential components of modeling, analysis, and problem-solving. Simplifications are the process of reducing complex systems or phenomena to simpler and more manageable forms, whereas assumptions are the premises or conditions that are considered to be valid for a specific analysis or model. In this section, we will investigate the significance of assumptions and simplifications, their function in various fields, and their potential influence on models' and analyses' applicability and validity. Because they aid in the development of mathematical or conceptual frameworks and simplify complex phenomena, assumptions are essential in modeling and analysis. It is common to assume that one should concentrate on the most important aspects of a system or problem and ignore the less important aspects. They make it possible to look into cause-and-effect relationships, make predictions, and make decisions by providing a starting point for analysis. Because it is often impossible to take into account every aspect of real-world systems, assumptions are necessary.

Depending on the context and objective of the analysis, assumptions can take many different forms. They might include presumptions regarding the behavior of variables, the relationships between variables, the absence of particular influences or interactions, or the truthfulness of particular laws or principles. Expert judgment, theoretical considerations, simplifying assumptions from previous studies, or empirical observations can all serve as foundations for assumptions. Improvements, then again, include reducing complex frameworks or peculiarities to less complex and more sensible structures. Models and analyses can be simplified to make them easier to understand, analyze, or solve. They include dismissing or approximating specific parts of a framework that are considered less pertinent or too complex to be in any way remembered for the examination. Improvements can be applied to different components of a framework, for example, the calculation, actual properties, limit conditions, or the elements of the framework.

The utilization of suppositions and disentanglements isn't without limits and expected traps. It is essential to acknowledge that assumptions are merely oversimplifications of the real world and may not be accurate in all circumstances. The context and scope of the analysis determine the assumptions' validity and applicability. Results and conclusions can be biased or unreliable if assumptions are made in the wrong way. Besides, improvements can present mistakes or vulnerabilities in the examination. Neglecting important factors or interactions that could have a significant impact on the outcomes may be part of simplifying a complex system. It is essential to evaluate the potential impact of simplifications on the analysis's accuracy and dependability. The robustness and limitations of simplified models and analyses can be evaluated with the assistance of sensitivity analyses or validation against experimental data.

The job of presumptions and rearrangements changes across various disciplines:

- **Investigative Studies:** Scientific research uses assumptions and simplifications to focus on particular aspects and simplify complex phenomena. In physics, for instance, it is common to practice making assumptions about idealized conditions or simplified mathematical models to make analysis easier and provide insight into fundamental principles. However, through empirical observations and experimental validations, scientists are constantly improving and challenging assumptions.
- **Design and engineering:** Suppositions and rearrangements are widely utilized in designing and configuration processes. To create models or designs, engineers make assumptions about the properties of materials, the conditions of the environment, or the behavior of systems. Rearrangements are applied to lessen computational intricacy, improve plans, or make designing estimations more reasonable. However, engineers need to carefully validate assumptions and think about how they will affect the safety and dependability of the final product or system.
- **Finance and economics:** Suppositions assume a huge part in monetary policy and monetary display. Suspicions about human behavior, economic situations, and level-headed navigation are frequently made to examine monetary peculiarities, gauge drifts,

or assess strategy influences. Economic models and frameworks are created using simplifications like rational expectations or assuming perfect competition.

Further Experimental Investigations

Further trial examinations are an urgent part of a logical exploration and designing a turn of events. Controlled experiments are used in experimental studies to gather data, test hypotheses, and learn more about a system's or phenomenon's behavior and properties. Through these investigations, theoretical predictions can be tested, prototype performance can be evaluated, and models or designs can be improved. We'll talk about why more experiments are needed, what to keep in mind when doing them, and how they can be used in a variety of fields in this section.

There are numerous reasons why additional experimental studies are necessary. To begin, they offer concrete evidence to either support or challenge theoretical models, hypotheses, or predictions. Researchers can verify the validity and accuracy of their findings by comparing their theoretical expectations with actual observations made with experimental data. Also, trial examinations can reveal startling peculiarities, irregularities, or limits that were not represented in the beginning models or speculations. These discoveries might prompt new experiences, updates on existing speculations, or the advancement of creative methodologies.

Experiments necessitate careful preparation and consideration of a variety of factors. To begin, the experimental setup needs to be constructed in such a way that it maintains the natural manifestation of the system or phenomenon under investigation while simultaneously controlling the relevant conditions and variables. To direct the experiment and guarantee that the data gathered is in line with the research objectives, clear objectives and research questions should be established. For data collection to be accurate, precise, and reproducible, proper protocols and procedures must be established.

The size of the sample and statistical analysis should also be taken into account. For statistically significant results and to lessen the impact of random variations, sufficient samples should be used. The data can then be analyzed, the significance of the findings evaluated, and reliable conclusions drawn using statistical methods. Moreover, tests ought to represent expected wellsprings of error or inclination and carry out controls or alleviation methodologies to limit their effect on the outcomes.

Trial examinations are key to logical exploration across disciplines like material science, science, and brain research. Experiments are used by scientists to test hypotheses, investigate natural phenomena, and either confirm or refute existing theories. In physics, for instance, experiments are carried out to examine the properties of materials, investigate the behavior of subatomic particles, or confirm the predictions made by brand-new theories. In biology, experimental studies aid in the comprehension of biological processes, the evaluation of drug efficacy, or the investigation of the effects of environmental factors on living things.

In engineering, experimental investigations are essential for assessing the safety, dependability, and performance of prototypes, materials, or systems. Engineers lead analyses to approve plans, advance boundaries, and evaluate the behavior of designs, machines, or gadgets. For instance, in aviation design, tests are performed to test the streamlined attributes of airplane models, survey the primary honesty of parts, or assess the proficiency of impetus frameworks. When it comes to testing materials, validating simulations, and optimizing design parameters, experimental investigations are also essential in fields like civil engineering, mechanical engineering, and electrical engineering.

Exploratory examinations are fundamental in clinical and medical care exploration to propel comprehension of illnesses, foster new therapies, and assess clinical mediations. Clinical preliminaries, a kind of trial examination, are directed to survey the viability and well-being of new medications or operations. In biomedical research, experimental studies aid in elucidating disease mechanisms, evaluating the efficacy of treatments, and examining the influence of lifestyle factors on health outcomes. Trial examinations in clinical imaging add to the improvement of symptomatic methods and imaging advances.

In environmental science, experimental studies are essential for evaluating the effects of climate change, pollution, and human activities on ecosystems. Scientists lead trials to look at the reaction of organic entities to ecological stressors, examine the proficiency of relief gauges, and evaluate the viability of reclamation endeavors. To simulate and comprehend intricate ecological processes, environmental science experimental investigations typically involve either field experiments, laboratory studies, or a combination of the two.

Advanced Modeling Techniques

The terms "advanced modeling techniques" and "methodologies" are used to describe the complex methods, phenomena, and systems that are used to accurately represent them. Advanced mathematical and computational tools are utilized in these methods, which go beyond straightforward analytical or empirical models to capture intricate relationships, dynamics, and behaviors. High-level demonstrating strategies are applied in different fields, including design, physical science, financial aspects, science, and sociology. In this passage, we will investigate the significance of cutting-edge display procedures, feature a few regularly utilized approaches, and examine their applications in various spaces. High-level displaying procedures assume a significant role in understanding and foreseeing complex frameworks that can't be sufficiently depicted by conventional models. Nonlinear relationships, emergent properties, feedback loops, and other intricate dynamics found in complex systems can be captured using these methods. Scientists and engineers can predict, optimize processes, and guide decision-making by simulating and analyzing the behavior of complex systems under various conditions using advanced modeling techniques.

The process of representing distinct entities, or agents, as well as their interactions within a system is known as agent-based modeling (ABM). Every specialist has its arrangement of rules, ways of behaving, and dynamic capacities. ABM permits analysts to concentrate on how the aggregate behavior of individual specialists leads to large-scale examples and peculiarities. It is especially helpful for concentrating on friendly frameworks, natural frameworks, traffic streams, and monetary business sectors. Computational Fluid Dynamics (CFD) is a strong displaying method for dissecting liquid streams, heat movements, and related peculiarities. CFD utilizes mathematical strategies to settle the administering conditions of liquid elements and capture complex stream ways of behaving. Aerospace engineering, automotive design, environmental engineering, and industrial processes all make extensive use of it.

Modeling the behavior and interactions of discrete entities and events over time is the basis of discrete event simulation (DES). DES records the order and timing of events as well as their effects on the system. It is generally utilized in tasks such as research, assembly, coordinated factors, and medical services to upgrade processes, assess framework execution, and pursue informed choices. The structure and function of biological neural networks serve as inspiration for a class of machine learning models called neural networks. They are fit for learning complex examples, making expectations, and performing errands in light of preparing information. Brain networks are utilized in different applications, for example, picture and discourse acknowledgment, normal language handling, monetary display, and prescient examination.

The goal of system dynamics modeling is to comprehend a system's feedback loops, delays, and nonlinear relationships. It underlines the investigation of framework conduct over the long haul and empowers scientists to mimic the drawn-out elements and results of strategy choices or interventions. Environmental modeling, public policy analysis, and the creation of business strategies all make use of system dynamics. In engineering fields like civil engineering, mechanical engineering, and electrical engineering, advanced modeling methods are used to simulate and improve complex systems. CFD is used, for instance, to evaluate structural dynamics in civil engineering projects, optimize thermal management in electronics, and analyze fluid flow in aerospace design.

In economics and finance, sophisticated modeling methods are frequently used to comprehend intricate market dynamics, forecast trends, and evaluate the effects of policy. A specialist-based demonstration is applied to concentrate on the way financial specialists behave, recreate economic situations, and examine the impacts of strategy mediations. Econometric models and brain networks are utilized for monetary estimation, risk appraisal, and venture systems. In healthcare and biomedical research, advanced modeling techniques are useful for studying disease progression, evaluating treatment options, and predicting outcomes.

CONCLUSION

In solid mechanics, the temperature dependence of flow stress is crucial. The versatility hypothesis gives a system for figuring out the disfigurement and stream conduct of materials under applied loads, permitting specialists and scientists to foresee and examine the way designs and parts behave. The phenomenon of a material's flow stress or resistance to plastic deformation varying with temperature is known as the temperature dependence of flow stress. Engineers can use plasticity theory to create and examine structures that can withstand external loads while maintaining performance and safety. It takes into account materials' plastic behavior, which is characterized by constant deformation without proportional stress increases. The concepts of yield criteria, strain hardening, and flow rules, which are necessary for accurately predicting the deformation and failure of materials under various loading conditions, are explained by plasticity theory. Engineers can use this theory to select the best materials and optimize the design of structures to guarantee their longevity and structural integrity.

The temperature dependence of stream pressure is a basic thought in strong mechanics, as it fundamentally impacts the deformity conductivity and mechanical properties of materials. There are three main regimes in which various materials exhibit distinct temperature-dependent flow stress behaviors: the low-temperature regime, in which the flow stress goes up as the temperature goes down; the intermediate-temperature regime, in which the flow stress stays pretty much the same; and the high-temperature regime, in which the flow stress goes down as the temperature goes up. For a variety of applications, it is essential to comprehend how flow stress is influenced by temperature. For instance, in assembling cycles, such as metal shaping, the temperature dependence of stream pressure determines the ideal temperature range for accomplishing the desired plastic deformity and lessening framing powers. To analyze the behavior of materials under various environmental conditions, such as elevated temperatures in fire scenarios or cryogenic temperatures in aerospace applications, the temperature dependence of flow stress is essential in structural engineering.

The material's crystal structure, microstructure, and the presence of impurities or alloying elements all have an impact on the temperature dependence of flow stress. At the smallest level, temperature influences the development of disengagements, which are the essential transporters of plastic misshapen Ing in glasslike materials. Dislocation mobility and interaction alter with temperature, resulting in variations in flow stress. To evaluate the temperature dependence of stream pressure, observational connections, for example, the Arrhenius condition or the power regulation condition, are usually utilized. These connections relate the stream pressure to temperature and other material boundaries. The flow stress-temperature relationship for specific materials is determined using experimental methods like hot deformation testing or compression testing at various temperatures. To simulate and predict the temperature-dependent behavior of materials, advanced modeling techniques like crystal plasticity modeling and finite element analysis are also utilized.

The temperature dependence of flow stress and plasticity theory are crucial components of solid mechanics. The versatility hypothesis gives an establishment for grasping the plastic deformity conduct of materials, empowering precise forecasts of the primary way of behaving and execution. Materials' deformation characteristics at various temperatures are affected by the temperature dependence of flow stress, which has an effect on their mechanical properties and behavior under various loading conditions. The comprehension of these ideas is fundamental for planning powerful designs, advancing assembly processes, and guaranteeing the dependability and well-being of designed frameworks. The field of solid mechanics and its applications in a variety of industries will continue to advance as a result of ongoing research and development in plasticity theory and the temperature dependence of flow stress.

Bibliography

1. Fung, Y. C. (2013). A Prologue to Continuum Mechanics Training from Pearson
2. The Mathematical Theory of Elasticity: Love, A. Edward H. Dover Distributions of a Treatise.
3. R. W. Ogden (1997) Disfigurements that aren't always easy to fix. Dispatch Organization.
4. J. Lubliner (2008) Hypothesis of Versatility Distributions of Dover.
5. Hill, R. (1950). The mathematical theory of plasticity. Oxford School Press.
6. R. W. Hertzberg (1996). Designing Materials' Break and Deformity Mechanics. John Wiley and Kids.
7. Jones, D. R. H., and M. F. Ashby Materials for Designing 1: Butterworth-Heinemann's design, applications, and properties in general.
8. Cook, N. G., and J. C. Jaeger Rock mechanics essentials John Wiley and Youngsters.
9. The Elasticity Theory by Goodier, J. N., and S. P. Timoshenko, McGraw-Hill.
10. A. F. Thicket (2010) Solids Applied Mechanics CRC Press.
11. Malvern Prologue to the Mechanics of a Constant Medium by E. Malvern. Prentice-Hall.
12. Elasticity's Mathematical Foundations, published by Dover by Hughes, T. J. R., and Marsden, J. E.
13. M. E. Gurtin (1981). An Introduction to Continuum Mechanics. The Academic Press.
14. J. R. the hairdresser Elasticity. Business Media and Springer Science
15. I. S. Sokolnikoff (2012) Mathematical Speculation of Flexibility. Dover Circulations.
16. The Elasticity Theory, by W. Nowacki (Springer Science and Business Media, 2005).

17. Biot, M. A. (1965). Gradual Distortions' Mechanics John Wiley and Kids.
18. The Limited Component Technique: O. C. Zienkiewicz, R. L. Taylor, and J. Z. Zhu Its foundation and fundamentals. Butterworth-Heinemann.
19. Reddy, J. N. (2013). A Quick Synopsis of the Cambridge University Press for Continuum Mechanics
20. P. Wriggers (2006) Techniques with Limited Component Nonlinearity Springer Science and Business Media.
21. A Prologue to the Mechanics of Solids by S. H. Crandall, N. C. Dahl, and T. J. Lardner McGraw-Hill.
22. G. T. Mase (2009) Architects' Guide to Continuum Mechanics CRC Press.
23. William S. Slaughter's Elasticity Linearized Theory Birkhäuser.
24. Advanced Materials Mechanics by A. P. Boresi, O. M. Sidebottom, and R. J. Schmidt John Wiley and Kids.
25. G. A. Maugin (1993). Thermomechanics of Plasticity and Fracture. Cambridge University Press
26. Hill, R. (1998). Liquid and strong mechanics. Cambridge University Press
27. S. H. Crandall, N. C. Dahl, and T. J. Lardner Mechanics of solids. Farewell McGraw-Incline Preparing.
28. Malvern An Overview of the Operations of a Continuous Medium Pearson Education, by E. Malvern.
29. Thermodynamics and mechanics of Continua by E. Fried and M. E. Gurtin Press of Cambridge College.
30. Digital Signal Processing and Modeling for Statistical Data, by M. Hayes (2014) John Wiley and Kids.
31. Budynas, R.G., Nisbett, J.K. Shigley's Mechanical Planning Plan. 2014, McGraw-Slope Schooling
32. Twisting and Crack Mechanics of Designing Materials, R.W. Hertzberg. 2012, John Wiley and Children
33. Goodier, J.N., and S.P. Timoshenko's Hypothesis of Flexibility McGraw-Hill Education, 1970.
34. Fung, Y.C. (2013), A Basic Course in Continuum Mechanics: Prentice Corridor
35. S.H. Crandall, N.C. Dahl, and T.J. Lardner wrote An Introduction to the Mechanics of Solids. 1999, McGraw-Slope Schooling.
36. R. Slope, Pliancy's Numerical Hypothesis. 1998, Oxford College Press.

37. Cottrell, A.H., Bilby, B.A., Swinden, K.H. Separations and Plastic Stream in Jewels. Clarendon Press
38. 2009. V.A. Lubarda's Elastoplasticity Theory, CRC Press, 2012.
39. Suiker, A.S.J. Mechanics of Adaptable Plans. 2009, CRC Press
40. Materials based on biology: N. Pugno Mechanics and Plan. 2017: Springer
41. Hypothesis of the Flexibility of an Anisotropic Versatile Body, by S.G. Lekhnitskii. 2013, Dispatch Partnership
42. Pinsky, P.M., 2013, Introduction to Continuum Mechanics, Cambridge University Press
43. Standards of Continuum Mechanics, J.N. Reddy: An Investigation of Reasonable Utilizations of Preservation Standards. 2008, Cambridge College Press
44. A.C. Eringen's Nonlinear Theory of Continuous Media. Dover Dispersions, 2012.
45. Malvern, L.E. Introduction to the Mechanics of a Consistent Medium. Prentice-Hall, 1969.
46. I.H. Shames and F.A. Cozzarelli conducted an analysis of elastic and inelastic stresses in the year 2003.
47. Academic Press 48's An Introduction to Continuum Mechanics, written by M.E. Gurtin in 2010.
48. Elasticity Not Linear: Theory and Implementation (2014), R.W. Ogden, CRC Press
49. Toll, T.C.T. Anisotropic Flexibility: Speculation and Applications. Oxford University Press, 1996.
50. R.S. Viscoelastic Solids, Lakes 2009, CRC Press
51. 2013 edition of A.E.H. Love's Treatise on the Mathematical Theory of Elasticity.
52. Dover Publications Lai, W.M., Rubin, D., Krempl, E. Preface to Continuum Mechanics. 2010, Elsevier
53. Lakes, R. Materials that are viscoelastic. Cambridge University Press.
54. 2009. Y.C. Fung Biomechanics: Living Tissues' Mechanical Properties Springer, 2013.
55. P.M. Naghdi 2012, Theory of Shells and Plates, Springer Science and Business Media
56. R.F. Gibson, 2013, Principles of Composite Material Mechanics, CRC Press
57. Luo, A.C.J. Principles of Composite Material Mechanics. 2018: CRC Press

58. C.O. Horgan, 2008, Finite Elements and Approximation, Cambridge University Press.
59. Choung, S.K., and Choung, J. Fatigue Crack Growth: 2011: Behavior, prediction, and mechanics Springer
60. Kim, J., Farris, T.N., Plant administrator, D.P. Micromechanics of Composite Materials: 2013 Elsevier
61. A Method for Generalized Multiscale Analysis Fung, Y. C. (2013). A First Course in Continuum Mechanics, fourth edition. Education from Pearson
62. Goodier, J. N., and S. P. Timoshenko Third ed., Hypothesis of Flexibility McGraw-Slope.
63. Elasticity's Mathematical Foundations, published by Dover by Hughes, T. J. R., and Marsden, J. E.
64. Love, A. E. H. (2013). A Treatise on the Mathematical Theory of Elasticity Dover Distributions, Fourth Edition.
65. E. Malvern, Malvern An Outline of the Operations of a Constant Medium Pearson Instruction.
66. R. W. Ogden (1997) Elastic or non-linear deformations Dover publications.
67. Hill, R. (1950). The mathematical theory of plasticity. Oxford School Press.
68. Limit design and plastic analysis of soil mechanics, or Drucker, D. C., and W. Prager Quarterly of Applied Mathematics, 10(2), 157-165.
69. (2009), Applied Mechanics of Solids by A. F. Bower CRC Press.
70. Solid Continuum Mechanics, CRC Press, 2007 by S. C. Cowin.
71. V. A. Lubarda (2010) Hypothesis of Elastoplasticity CRC Press.
72. A Brief Overview of Continuum Mechanics, published by Scholastic Press in 2008 by M. E. Gurtin.
73. Energy and finite element methods are utilized in structural mechanics, according to Dym, C. L., and Shames, I. H. SI numbers. CRC Press.
74. J. N. Reddy is the author. The Method of the Finite Element: A Summary (Third Edition) McGraw-Hill.
75. M. A. Crisfield (1997) Wiley's Non-linear Finite Element Analysis of Solids and Structures, Second Edition.
76. T. Belytschko, W. K. Liu, and B. Moran Continua and designs can profit from nonlinear limited components. Wiley.
77. K. J. Wash (1996). Second Edition of Limited Component Methods Pearson Schooling.

78. O. C. Zienkiewicz, R. L. Taylor, and others The 6th Release of the Limited Component Strategy for Strong and Underlying Mechanics Butterworth-Heinemann.
79. M. Pinsky, Pinsky, Limited and Ghostly Component Techniques Using MATLAB: A Preface (second edition), Springer.
80. J. Bonet, R. D. Wood, and others Nonlinear Continuum Mechanics for Restricted Part Assessment. Cambridge University Press
81. Biot, M. A. (1962). Steady Distortions' Mechanics Wiley.
82. Viscoelastic Materials, R. S. Lakes (1998) Cambridge University Press
83. Kachanov, L. M. (1992). An Overview of Continuum Damage's Mechanisms Martinus Nijhoff is the publisher.
84. Wierzbicki, T., and W. Abramowicz on the devastating mechanics of designs with meager walls. 72(5): Journal of Applied Mechanics, pp. 727–734.
85. Cell Solids: Ashby, M. F., and Gibson, L. J. Second version of Construction and Properties Press of Cambridge College.
86. Froth designs with a negative Poisson's proportion, by R. S. Lakes (2009). Science, 235(4792), 1038-1040.
87. Hill, R. (1952). the fundamental structure of polycrystalline constitutive laws and metal composites. Journal of the Mechanics and Physics of Solids, 1(1), 19–30.
88. Mill operator, M. F. Ashby Materials Decision in Mechanical Arrangement (fourth ed.). Butterworth-Heinemann.
89. S. Suresh (1998). Materials' Weariness, Second Release Press of Cambridge College.
90. The creator is J. W. Hutchinson. Break mechanics models: a few thoughts of my own. 138(1-4), 1-21, Worldwide Diary of Break.
91. S. Timoshenko, J. N. Goodier, and Theory of adaptability. McGraw-Hill.
92. Fung, Y. C. (1993). Biomechanics: Mechanical properties of living tissues. Springer Science and Business Media.
93. Hughes, T. J. R., and Marsden, J. E. Mathematical foundations of flexibility. Education from Pearson
94. Love, A. Edward H. a treatise on the mathematical theory of elasticity. Cambridge University Press
95. Ogden, R. W. (1997). deformations that are not linear but elastic. Courier is a company.
96. Hill, R. (1998). The mathematical theory of plasticity Oxford School Press.

97. W. Johnson (1985). The adaptable idea of metals: a book on metal plastic deformation. Cambridge University Press
98. (1996) R. W. Hertzberg Mechanical engineering of material deformation and fracture. John Wiley and Kids.
99. An overview of the operation of a continuous medium by E. Malvern, Malvern Education from Pearson
100. M. E. Gurtin (1981). Scholastic Press, a primer on continuum mechanics.
101. Pliancy hypothesis, by J. Lubliner (2008) Dover Circulations.
102. Hughes, T. J. R., and J. C. Simo (1998) Computing inelasticity. Business Media and Springer Science
103. Cottrell, A. H. (1981). an introduction to metallurgy. Butterworth-Heinemann.
104. Zerna, William, and A. E. Green Flexibility in principle. Oxford School Press.
105. Elasticity, J. R. Barber (2002). Scholarly Distributors, Kluwer.
106. Bacon, David J., and E. H. Dill Preface to the mechanics of a relentless medium. Pearson Instruction.
107. George A. Holzapfel, Solid nonlinear mechanics: a continuum-based approach to engineering. John Wiley and Kids.
108. A. Meguid, S. A. Materials mechanics: an education-integrated system. CRC Press.
109. Gere, J. M., and S. P. Timoshenko materials mechanics. Cengage Learning.
110. William S. Slaughter's linearized elasticity theory Springer Science and Business Media.
111. R. S. Lakes (1998) Solids that are thick. CRC Press.
112. Lakes, R. S. (1999). CRC Press's methods for experimenting with fracture mechanics
113. W. Prager (2004) Preface to mechanics of continua. Dover Circulations.
114. Kachanov, L. M. (1992). An outline of the harm mechanics of the continuum. Distributers Martinus Nijhoff.
115. The introduction to micro and nanomechanics by Ting, T. C. T., and Y. H. Pao The Science World
116. O. C. Zienkiewicz, R. L. Taylor, and others The technique for limited components: its foundation and fundamentals. Elsevier.
117. The author is J. N. Reddy. a brief introduction to continuum mechanics Cambridge University Press.

118. D. Krajcinovic, 1996, Damage mechanisms. North-Holland.
119. Wu, J. (2013). Thermodynamics and continuum mechanics: relating fundamental ideas to governing equations. Cambridge University Press
120. Preface to the flexibility hypothesis for gem absconds by Francis A. Leckie. Noordhoff.
121. Hill, R. (1963). Built up solids' versatile properties include: a few basic ideas in theory. 11(5): Journal of the Mechanics and Physics of Solids, volume 357–372.
122. P. M. Naghdi (1972). The idea of plates and shells. 12 pages of advances in applied mechanics 93-186). The Academic Press.
123. Microstructure of martensite: K. Bhattacharya (2003a) What it is and how and why it forms the shape-memory effect. Oxford School Press.
124. The authors are L. Cedolin and Z. P. Baant. Design dependability: speculations of damage, inelasticity, crack, and versatility. Dover publications.
125. G. Duvaut and J. L. Lions (1996) inequalities in mechanics and physics. Springer Science and Business Media.
126. Hill, R. (1998). Strong composites' flexible properties include: some fundamental ideas. Journal of the Mechanics and Actual study of Solids, 11(5), 357-372.
127. Oxford College Press, Principles of Dislocation Theory, by J. Weertman in 1996.
128. in discrete dislocations: Needleman, A., and E. van der Giessen A direct planar model. Showing and amusement in materials science and planning, 3(5), 689.
129. A. P. Boresi and Richard J. Schmidt Material mechanics at a high level. John Wiley and Kids.
130. Lin, T. Y., and Kinzel, G. L. (2002). Engineering mechanics: Elements and statics. Oxford School Press.
131. The limited component strategy: O. C. Zienkiewicz, R. L. Taylor, and J. Z. Zhu its foundation and fundamentals. Butterworth-Heinemann.
132. R. Sivasubramanian, S. Dharmaraja, and Rudiments of solid mechanics. CRC Press.
133. M. N. Sadiku (2011) electromagnetic components. Oxford School Press.
134. I. H. Shames and F. A. Cozzarelli conducted an investigation into flexible and inelastic pressure. CRC Press.

135. Solids mechanics by S. Rajasekaran and G. Sankarasubramanian. Private PHI Education Ltd.
136. Hill, R. (1990). Elastoplastic polycrystal continuum miniature mechanics 38(3): 405–421, Solid Mechanics and Physics Journal.
137. M. Petyt's (2013) introduction to the limited component vibration study. Cambridge University Press
138. The elements of multibody frameworks, by A. N. Guz and S. S. Antonov. Business Media and Springer Science
139. I. Elishakoff and D. Pentaras are the creators. Hypothesis of bifurcation applied: Strategies for discretization Springer Science and Business Media.
140. Zohdi, T. I. (2008). a first step toward showing and reproducing particulate streams. SIAM.
141. McGraw-Hill, S. Timoshenko, J. N. Goodier, and the elasticity theory.
142. Fung, Y. C. (1993). Biomechanics: Mechanical properties of living tissues. Business Media and Springer Science
143. The mathematical foundations of versatility are provided by Hughes, T. J. R., and J. E. Marsden. Education from Pearson
144. Love, A. Edward H. a treatise on the mathematical theory of elasticity. Cambridge University Press
145. R. W. Ogden (1997) Flexible misshapenings that are not straight. Courier is a company.
146. Hill, R. (1998). The mathematical theory of plasticity Oxford School Press.
147. W. Johnson (1985). The malleability of metals: a book on metal plastic deformation. Cambridge University Press
148. R. W. Hertzberg (1996). Designing materials' crack and deformity mechanics. John Wiley and Kids.
149. An overview of the operation of a continuous medium by E. Malvern, Malvern Education from Pearson
150. M. Edward Gurtin an introduction on continuum mechanics Educational Press.
151. J. Lubliner (2008) The hypothesis of versatility Distributions of Dover.
152. Simo computational inelasticity was developed by Hughes, T. J. R., and J. C. Business Media and Springer Science
153. Cottrell, A. H. (1981). an introduction to metallurgy. Butterworth-Heinemann.

154. W. Zerna, A. E. Green, and the theory of elasticity. Oxford School Press.
155. Elasticity, J. R. Barber (2002). Distributers Kluwer Scholarly
156. Bacon, D. J., and Dill, E. H. An outline of how a constant medium functions. Education from Pearson
157. Holzapfel, George A. Nonlinear mechanics that work: A continuum approach for planning. John Wiley and Kids.
158. Materials mechanics: A. Meguid, S. A. an education-integrated system. CRC Press.
159. Gere, J. M., and S. P. Timoshenko materials mechanics. Cengage Learning.
160. William S. Slaughter's linearized elasticity theory Business Media and Springer Science
161. R. S. Lakes (1998) Viscoelastic solids. CRC Press.
162. Lakes, R. S. (1999). Crack mechanics testing methods. CRC Press.
163. W. Prager (2004) A prologue to consistent mechanics. Dover publications.
164. Kachanov, L. M. (1992). An outline of the harm mechanics of the continuum. Martinus Nijhoff Distributers.
165. Chime, T. C. T., and Y. H. Pao A prologue to nano-and micromechanics. Globally logical.
166. The limited component technique: O. C. Zienkiewicz, R. L. Taylor, and others its fundamentals and premise. Elsevier.
167. The author is J. N. Reddy. a brief introduction to continuum mechanics Cambridge University Press.
168. Krajcinovic, D. (1996). Mechanisms for causing harm North-Holland.
169. Wu, J. (2013). Continuum mechanics and thermodynamics: relating fundamental ideas to governing equations. Press of Cambridge College.
170. An introduction to the theory of crystal defect elasticity by Francis A. Leckie. Noordhoff.
171. Hill, R. (1963). Supported solids' flexible properties include: a couple of hypothetical fundamentals. 11(5): 357-372, Diary of the Mechanics and Material science of Solids.
172. P. M. Naghdi's work the idea that plates and shells exist. 12 pages of advances in applied mechanics 93-186). Academic Press
173. Martensite's microstructure, as described by K. Bhattacharya (2003a): How and why it structures and results in the shape-memory effect. Oxford School Press.

174. The authors are L. Cedolin and Z. P. Baant. Structure security: Speculations of harm, inelasticity, crack, etc. Distributions by Dover.
175. Inequalities in physics and mechanics by G. Duvaut, J. L. Lions, and others. Springer Science and Business Media.
176. Hill, R. (1998). Elastic properties of solid composites include: a couple of key ideas. 11(5): 357-372, Diary of the Mechanics and Material science of Solids.
177. Oxford College Press, Principles of Dislocation Theory, by J. Weertman in 1996.
178. van der Giessen, E., and Needleman, A. (1995). Discrete separation flexibility: simple planar model for modeling and simulation in materials science and engineering, 3(5), 689.
179. Material mechanics were advanced by R. J. Schmidt and A. P. Boresi. John Wiley and Kids.
180. Kinzel, G. L., and T. Y. Lin. Engineering mechanics: Elements and statics Oxford School Press.
181. J. Z. Zhu, R. L. Taylor, and O. C. Zienkiewicz The finite element method: Its reason and essentials. Butterworth-Heinemann.
182. R. Sivasubramanian, S. Dharmaraja, and strong mechanics essentials CRC Press.
183. Electromagnetic elements, by M. N. Sadiku (2011) Oxford School Press.
184. Cozzarelli, F. A., and I. H. Disgraces Investigations of both flexible and inelastic pressure. CRC Press.
185. Solid mechanics by S. Rajasekaran and G. Sankarasubramanian Private PHI Education Ltd.
186. Hill, R. (1990). the continuum's micromechanics in elastoplastic polycrystals. 38(3): 405–421, Solid Mechanics and Physics Journal.
187. An introduction to vibration analysis using finite elements by M. Petyt (2013) Cambridge University Press
188. Guz, A. N., and S. S. Antonov Components of multibody systems. Springer Science and Business Media.
189. I. Elishakoff and D. Pentaras are the authors. Hypothesis of bifurcation applied: Springer Science and Business Media's discretization techniques
190. An introduction to modeling and simulating particulate flows by T. I. Zohdi (2008) SIAM.
191. Timoshenko, S., and Goodier, J. N. (1970). The flexibility hypothesis McGraw-Slope.

192. Y. C. Fung (1993). Biomechanics: Mechanical properties of living tissues. Business Media and Springer Science
193. Hughes, T. J. R., and J. E. Marsden Versatility's numerical establishments Pearson Instruction.
194. Love, A. Edward H. a composition on the versatility numerical hypothesis. Cambridge University Press
195. R. W. Ogden (1997) Flexible misshapenings that are not straight. Organization for Dispatch.
196. Hill, R. (1998). The mathematical speculation of flexibility. Oxford School Press.
197. Johnson, W. (1985). The plethora of uses for metals: a textbook on metal distortion within plastic reach. Cambridge University Press
198. (1996) R. W. Hertzberg Material design distortion and crack mechanics. John Wiley and Kids.
199. Malvern Prologue to the mechanics of a continuous medium by E. Malvern. Pearson Training.
200. M. Edward Gurtin Scholarly Press, a primer on continuum mechanics.
201. Versatility hypothesis, J. Lubliner (2008) Dover publications.
202. Hughes, T. J. R., and J. C. Simo Inelasticity in registering. Business Media and Springer Science
203. Cottrell, A. H. (1981). an introduction to metallurgy. Butterworth-Heinemann.
204. Hypothetical flexibility, W. Zerna, A. E. Green, and Oxford School Press.
205. Elasticity, J. R. Barber (2002). Distributers Kluwer Scholarly
206. An overview of the operation of a continuous medium, by D. J. Bacon and E. H. Dill. Education from Pearson
207. George A. Holzapfel, Nonlinear strong mechanics: a continuum-based approach to engineering. John Wiley and Kids.
208. A. Meguid, S. A. Mechanics of materials: an education-integrated system. CRC Press.
209. Materials mechanics by S. P. Timoshenko and J. M. Gere. Cengage Learning.
210. William S. Slaughter's linearized elasticity theory Business Media and Springer Science
211. R. S. Lakes (1998) Solids that are gooey. CRC Press.
212. R. S. Lakes (1999). CRC Press's methods for experimenting with fracture mechanics

213. An introduction to continuous mechanics by W. Prager (2004) Distributions of Dover.
214. Kachanov, L. M. (1992). Introduction to continuum damage mechanics Distributors Martinus Nijhoff
215. An introduction to nano- and micromechanics by Ting, Y. H. Pao, and T. C. T. Ting. Globally logical.
216. The finite element method: R. L. Taylor, O. C. Zienkiewicz, and others its foundation and fundamentals. Elsevier.
217. The creator is J. N. Reddy. a brief introduction to continuum mechanics Cambridge University Press.
218. D. Krajcinovic, 1996 Mechanical problems. North-Holland.
219. Wu, J. (2013). Continuum mechanics and thermodynamics: relating fundamental ideas to governing equations. Cambridge University Press
220. An introduction to the theory of crystal defect elasticity by Francis A. Leckie. Noordhoff.
221. Hill, R. (1963). Supported solids' flexible properties include: A couple of theoretical principles. Journal of the Mechanics and Material study of Solids, 11(5), 357-372.
222. P. M. Naghdi's work The idea of plates and shells. 12 pages of advances in applied mechanics 93-186). Press of Scholastics.
223. K. Bhattacharya (2003a) Martensite's microstructure: What it is and how and why it forms the shape-memory effect. Oxford School Press.
224. The authors are L. Cedolin and Z. P. Baant. Structure security: theories regarding damage, inelasticity, fracture, and other phenomena. Dover publications.
225. G. Duvaut, J. L. Lions, and others material science and mechanics disparities. Business Media and Springer Science
226. Hill, R. (1998). Elastic properties of solid composites include: some fundamental ideas. 11(5): Journal of the Mechanics and Physics of Solids, volume 357–372.
227. J. Weertman (1996) Standards of separation hypothesis Oxford School Press.
228. Needleman, A., and E. van der Giessen Flexibility in discrete disengagement: simple planar model for modeling and simulation in materials science and engineering, 3(5), 689.
229. Material mechanics were advanced by R. J. Schmidt and A. P. Boresi. John Wiley and Kids.

230. Lin, T. Y., and Glenn L. Kinzel Mechanics in designing: Elements and statics Oxford School Press.
231. J. Z. Zhu, R. L. Taylor, and O. C. Zienkiewicz The finite element method: its basics and establishment. Butterworth-Heinemann.
232. Dharmaraja, S., and R. Sivasubramanian CRC Press provides solid mechanics fundamentals.
233. M. N. Sadiku (2011) electromagnetic components Oxford School Press.
234. Analysis of both elastic and inelastic stress by I. H. Shames and F. A. Cozzarelli. CRC Press.
235. Sankarasubramanian, G., and Rajasekaran, S. mechanics of solids. Private PHI Education Ltd.
236. Hill, R. (1990). the continuum's micromechanics in elastoplastic polycrystals. Journal of the Mechanics and Material study of Solids, 38(3), 405-421.
237. M. Petyt's (2013) introduction to the examination of limited component vibration. Cambridge University Press
238. The elements of multibody frameworks, by A. N. Guz and S. S. Antonov. Business Media and Springer Science
239. I. Elishakoff and D. Pentaras are the creators. Application of bifurcation theory: Discretization methods Business Media and Springer Science
240. A preface to displaying and reenacting particulate streams, by T. I. Zohdi (2008) SIAM.
241. Hibbeler, R. C. (2011). 9th edition of Materials Mechanics Pearson.
242. S. P. Timoshenko and Jonathan N. Goodier The Flexibility Hypothesis (Third Edition) McGraw-Hill.
243. An Introduction to Continuum Mechanics, Second Edition, by M. E. Gurtin, Academic Press of Academics, 2010.
244. Y. C. Fung (2013) Fourth release of A First Course in Continuum Mechanics. Pearson Training.
245. A. E. H. Love (1944). The Elasticity Mathematical Theory: Dover Publications' A Treatise
246. Malvern, L. E. (1969). An Outline of the Functions of a Persistent Medium Prentice-Corridor.
247. J. Lubliner (2008) Flexibility Speculation (Upgraded Delivery). Distributions by Dover.
248. Hill, R. (1998). The mathematical theory of plasticity. Oxford School Press.

249. R. W. Ogden (1997) Flexible disfigurements that are not straight. Dover Dispersions.
250. The authors are E. H. Davis and H. G. Poulos. The Pile Foundation's design and investigation Wiley.
251. Rao, S. S. (2010). Pearson's Mechanical Vibrations, Fifth Edition.
252. Engineering Mechanics of Solids, second edition, by E. P. Popov, Pearson Education, 2017.
253. S. Suresh (1998). Materials' Fatigue, Cambridge University Press, Second Edition
254. Shames, I. H., and Cozzarelli, F. A. (1997). The Second Edition of the Versatile and Inelastic Pressure Examination Education from Pearson
255. An Introduction to the Mechanics of Solids, Second Edition, by S. H. Crandall, N. C. Dahl, and T. J. Lardner, McGraw-Hill.
256. Elasticity: N. J. Pagano and P. C. Chou Dyadic, tensor, and designing techniques Dover publications.
257. R. Lakes (1999). Solids that are viscous. CRC Press.
258. (2013a) T. Roubek Nonlinear Partial Differential Equations and Their Applications (Third Edition). Springer.
259. Hughes, T. J. R., and J. C. Simo Calculated inelasticity. Springer.
260. Prologue to Designing Pliancy by R. K. N. D. Rajapakse (2011) Elsevier.

Made in the USA
Columbia, SC
16 September 2026

cb6fe98d-b424-48c1-9b59-a9b9b27b4e48R01